Biotechnology Principles

John Smith is Professor of Applied Microbiology at the University of Strathclyde in the new Department of Bioscience and Biotechnology. A leading international authority on biotechnology, his research interests include fungal physiology and biochemistry, mycotoxicology and solid substrate fermentations.

Aspects of Microbiology 11

Biotechnology Principles

John E. Smith

Van Nostrand Reinhold (UK) Co. Ltd

For Jamie

First published in 1985 by
Van Nostrand Reinhold (UK) Co. Ltd
Molly Millars Lane, Wokingham, Berkshire, England

Reprinted 1985

Photoset in Times 9 on 10pt by Kelly Typesetting Limited
Bradford-on-Avon, Wiltshire

Printed and bound in Hong Kong

ISBN 0 442 30590 7
(ISSN 0266–6642)

Preface

Biotechnology concerns applied bioscience and technology and involves the practical application of biological organisms, or their subcellular components, to manufacturing and service industries and to environmental management. Successful application of biotechnology can only be achieved from the integration of several scientific disciplines and technologies including biochemistry, microbiology, genetics and molecular biology and chemical and process engineering. This small book attempts to bring together the basic principles around which biotechnology has developed, and it is hoped that it will allow students or practitioners in the main component disciplines to obtain an integrated overview of biotechnology.

JOHN E. SMITH, *Glasgow*

Contents

1 Introduction

The nature of biotechnology

Biotechnology is an area of applied bioscience and technology which involves the practical application of biological organisms, or their subcellular components to manufacturing and service industries and to environmental management. Biotechnology utilizes bacteria, yeasts, fungi, algae, plant cells or cultured mammalian cells as constituents of industrial processes. Successful application of biotechnology will result only from the integration of a multiplicity of scientific disciplines and technologies, including microbiology, biochemistry, genetics, molecular biology, chemistry and chemical and process engineering.

Biotechnological processes will normally involve the production of cells or biomass, and the achievement of desired chemical transformations. The latter may be further subdivided into:

(a) formation of a desired end product (e.g. enzymes, antibiotics, organic acids, steroids);
(b) decomposition of a given starting material (e.g. sewage disposal, destruction of industrial wastes or oil spillages).

The reactions of biotechnological processes can be catabolic, in which complex compounds are broken down to simpler ones (glucose to ethanol), or anabolic or biosynthetic, whereby simple molecules are built up into more complex ones (antibiotic synthesis). Catabolic reactions are usually exergonic whereas anabolic reactions are normally endergonic.

Biotechnology includes fermentation processes (ranging from beers and wines to bread, cheese, antibiotics and vaccines), water and waste treatment, parts of food technology, and an increasing range of novel applications ranging from biomedical to metal recovery from low grade ores. Because of its versatility, biotechnology will exert a major impact in many industrial processes and in theory almost all organic materials could be produced by biotechnological methods. Predictions of future worldwide market potential for biotechnological products in the year 2000 have been estimated at nearly US $65bn (Table 1.) However, it must also be appreciated that many important new bio-products will still be synthesized chemically from models derived from existing biological molecules, e.g., new drugs based on the interferons. Thus the interface between bioscience and chemistry and its relationship to biotechnology must be broadly interpreted.

A high proportion of the techniques used in biotechnology tend to be more economic, less energy demanding and safer than current traditional industrial processes and for most processes the residues are biodegradable and non-toxic. In the long term biotechnology offers a means of solving some major world problems, in particular those related to medicine, food production, pollution control and the development of new energy sources.

Table 1 Growth potential for worldwide biotechnological markets by the year 2000

Market sector	$ (million)
Energy	16 350
Foods	12 655
Chemicals	10 550
Health care (pharmaceuticals)	9080
Agriculture	8546
Metal recovery	4570
Pollution control	100
Other (i.e. unexpected developments)	3000
Total	64 851

From T. A. Sheets Co. (1983). *Biotechnology Bulletin* November.

Historical evolution of biotechnology

Contrary to popular belief biotechnology is not a new pursuit but in reality dates far back into history. In practice, four major developmental phases can be identified in arriving at modern biotechnological systems

1 Biotechnological production of foods and beverages Activities such as baking, brewing and wine making are known to date back several millenia; the ancient Sumarians and Babylonians were drinking beer by 6000 B.C., the Egyptians were baking leavened bread by 4000 B.C. while wine was known in the Near East by the time of the book of Genesis. The recognition that these processes were being affected by living organisms, yeasts, was not formulated until the 17th century, by Anton van Leeuwenhoek. Definitive proof of the fermentative abilities of these minute organisms came from the seminal studies of Pasteur between 1857 and 1876. Pasteur can justifiably be considered as the father of biotechnology.

Other microbially based processes such as the production of fermented milk products, e.g. cheeses and yogurts, and various oriental foods, e.g. soy sauce, tempeh etc., can equally claim distant ancestry. Of more recent introduction is mushroom cultivation which probably dates back many hundreds of years for Japanese shii-ta-ke cultivation and about 300 years for the *Agaricus* mushroom now widely cultivated throughout the temperate world.

It cannot be ascertained whether these microbial processes arose by accidental observation or by intuitive experimentation but their further and continued development were early examples of man's abilities to use the vital activities of organisms for his own needs. In more recent times, just as these processes have become more reliant on advanced technology, their contribution to world economy has equally increased out of all proportion to their humble origins.

2 Biotechnological processes initially developed under non-sterile conditions Many important industrial compounds such as ethanol, acetic acid, organic acids, butanol and acetone were being produced by the end of the 19th century by microbial fermentation procedures that were open to the environment; the control of contaminating micro-organisms was achieved by careful manipulation of the ecological environment and *not* by complicated engineering practices.

However, with the arrival of the petroleum age in which these compounds could be produced more cheaply from the by-products of petroleum production, most of these emerging industries were eclipsed. Escalation of oil prices in recent years has lead to a re-examination of the early fermentation procedures with a view to a return to commercial production. Together with the previously mentioned food fermentations these fermentation practices are relatively simple and can be run on a large scale.

Other outstanding examples of non-sterile biotechnology are waste water treatment and municipal composting of solid wastes. Microorganisms have long been exploited for purposes of decomposing and detoxifying human sewage and, to a lesser extent, in the treatment of industrial toxic wastes such as those from the chemicals industry. Biotechnological treatment of waste waters represents by far the largest (but least recognised) fermentation capacity practised throughout the world (Table 2).

3 The introduction of sterility to biotechnological processes A new direction in biotechnology came in the 1940s with the introduction of complicated engineering techniques to the mass cultivation of micro-organisms to ensure that the particular biological process could proceed with the exclusion of contaminating micro-organisms. Thus by prior sterilization of the medium and the bioreactor and with engineering provision for the exclusion of incoming contaminants only the chosen biocatalyst was present in the reactor. Examples of such products, which represent an increasing volume of biotechnological activity, include antibiotics, amino acids, organic acids, enzymes, steroids, polysaccharides and vaccines. Most of these processes are complicated, expensive and suitable only for high value products. Although many of the products are produced in relatively large quantities they are still dwarfed in volume and financial returns by the older systems used in food and beverage biotechnology (Table 3).

4 New dimensions and possibilities for biotechnological industries Within the last decade there have been outstanding developments in molecular biology and process control which have created new and exciting opportunities not only to create new dimensions but also to improve greatly the efficiency and economics of the established biotechnological industries. It is largely from these discoveries and developments that there have been such euphoric statements about the future role of biotechnology to the world economy.

What then are these new innovations? (Table 4).

Table 2 Total UK fermentation capacity

Product	Total capacity (m^3)
Waste water	2 800 000
Beer	128 000
Baker's yeast	19 000
Antibiotics	10 000
Cheese	3000
Bread	700

From Dunnill (1981).

Table 3 UK fermentation industry sales

Industry	Sales (£m)
Brewing	3190
Spirits	1860
Cheese	415
Cider, wine	190
Bread	150
Antibiotics	100
Yogurt	65
Yeast	25
Citric acid	20

From Dunnill (1981).

(a) Genetic engineering. Manipulation of the genome of industrially important organisms by sexual recombination and/or by mutation have long been part of the innovative repertoire of the industrial geneticist. New recombinant DNA techniques involve breaking living cells gently, the extraction of DNA, its purification and subsequent selective fragmentation by highly specific enzymes; the sorting, analysis, selection and purification of a fragment containing a required gene; chemical bonding to the DNA of a carrier molecule and the introduction of the hybrid DNA into a selected cell for reproduction and cellular synthesis. Recombinant DNA technologies permit easier manipulation of a genome and can readily bypass interspecies and intergeneric incompatability. Unlimited possibilities exist and already human insulin and interferon genes have been transferred into and expressed by microbial cells. Protoplast fusion, monoclonal antibody preparation and the wide use of tissue culture techniques including the regeneration of plants from suspension culture cells have had a profound impact on the development of biotechnology. (Chapter 2).

(b) Enzyme technology. Isolated enzymes have long been a part of many biotechnological processes and their catalytic properties are being further utilized with the development of suitable immobilization techniques allowing reuse of the biocatalyst. Of particular importance has been the development of high fructose syrups (annual production 3 million tonnes) using immobilized bacterial glucose

Table 4 Techniques stimulating the development of biotechnology

Recombinant DNA manipulation
Tissue culture
Protoplast fusion
Monoclonal antibody preparation
Protein structural modification ('protein engineering')
Immobilized enzyme and cell catalysis
Sensing with the aid of biological molecules
Computer linkage of reactors and processes
New biocatalytic reactor design

After Dunnill & Rudd (1984).

isomerase. A further development is the immobilization of whole cells for biocatalytic purposes (Chapter 4).

(c) Biochemical engineering. Bioreactors play a central role in biotechnological processes by providing a link between the starting materials or substrates and final products (Figure 1; Chapter 3). Major advances have been made in bioreactor designs, in process monitoring techniques and in computer control of fermentation processes. However, the application of process control in biotechnological industries is many years behind that in operation in the chemical process industry. New approaches to the processing of the products of biotechnology (downstream processing) will improve the economics of all processes. There is an increasing need to design efficient recovery processes, in particular for high value products e.g. the ratio of recovery to fermentation costs for L-asparaginase is about 3.0 whereas for ethanol it is 0.16 (Chapter 5). However, downstream processing is still the Cinderella subject of biotechnology.

(d) Engineered products/systems. The ability to produce in quantity biological molecules such as antibodies or enzymes together with the techniques of protein and cell immobilization are allowing the development of radically new sensors that can be used for biodiagnostic, and biodetoxification. purposes. Such systems can be combined with micro-electronic devices and ultimately computers allowing sophisticated control programming in many biotechnological industries and services.

Biotechnology has two characteristic features: its connection with practical applications, and interdisciplinary cooperation. The practitioners of biotechnology will employ techniques derived from chemistry, microbiology, genetics, biochemistry, chemical engineering and computer science, and their main mission will be the innovation, development and optimal operation of processes in which biochemical catalysis has a fundamental and irreplaceable role. Biotechnology does not constitute a new discipline but rather is an activity in which specialists from a wide range of disciplines can make their contributions.

A clear distinction must be drawn between bioscience and biotechnology. Bioscience refers to acquisition of biological knowledge whereas biotechnology

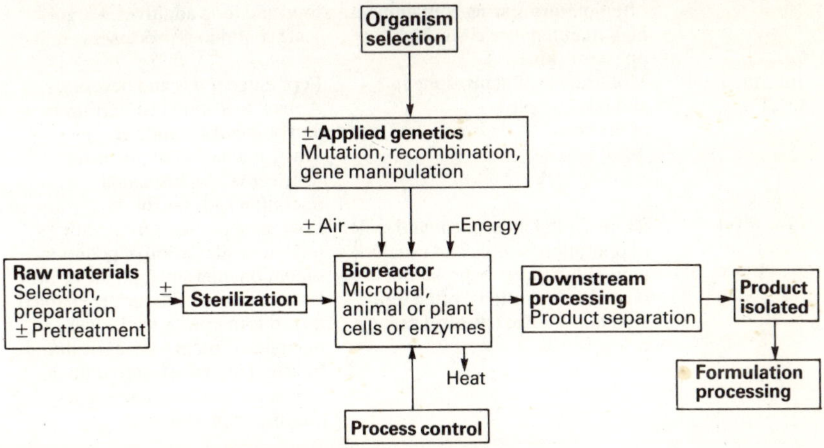

Fig. 1 Schematic overview of a biotechnological process.

refers to application of biological knowledge. Biotechnological processes will, in most cases, function at low temperatures, will consume little energy and rely in general on inexpensive raw materials as substrates.

Bioscientists and engineers of various specializations will make their individual contributions to biotechnology. The term *biotechnologist* has crept into our vocabulary as an all-embracing description of scientists or engineers engaged in applying their skills and knowledge to the *processing* of biological materials. However, the use of this term should be discontinued as it can only lead to confusion. In contrast a *biochemical engineer* is a process engineer whose role is to *transfer* the knowledge of the biological scientist into a practical operation. A biochemical engineer will have been trained in the scientific and engineering principles underlying the design and operation of biological operations.

A 'complete biotechnologist' will never exist, since no one can be expert in the skills of microbiology, biochemistry, molecular biology, chemical and process engineering, etc. However, for those who practise in this subject every effort must be made to understand the language of the other component subjects. The lack of a common language between specialists in different disciplines is undoubtedly the major obstacle in realising the full potential of biotechnology.

Application of biotechnology

Biotechnological processes can be considered on the basis of volume and value. Thus, high volume, low value products or services include water purification and

Table 5 Biotechnology: based on technological level

Category	Input	Output
High level	High capital investment; sophisticated plant and processes often requiring strict containment; high maintenance costs; high operator skills.	High value-added products destined for health care and human food and food additives. Large-scale continuous processes.
Intermediate level	Moderate capital investment and less complex operations.	Fermented foods and beverages. Animal feedstuffs. Biofertilizers and pesticides. Crude enzymes. Waste management processes which entail sophisticated operation and control.
Low level	Small capital investment and scale of operation, simple and usually indigenous equipment; labour intensive operations; often septic systems. Village level technology.	Low value products frequently related to alleviation of pollution, sanitation, fuel and food provision. Extensive use of naturally adapted mixed fermentations. Biogas; microbial protein from agricultural food and food wastes; traditional fermented foods and beverages; mushroom production.

From Bull *et al*. (1982).

effluent and waste treatment and the production of methane, ethanol, biomass and animal feed; relatively high volume, intermediate value products include amino acids and organic acids, food products, baker's yeast, acetone, butanol and certain polymers, while low volume, high value products include antibiotics, interferons, vaccines, monoclonals, antibodies, enzymes and vitamins. Biotechnology can also be considered in terms of the levels of technology that will be necessary for product formation (Table 5).

With reference to the scale of industrial development rather than the size of the individual production units, present and future biotechnology can also be conveniently divided into three areas.

(a) *Small scale biotechnology* is specifically concerned with biochemical products that can be produced economically only by biological means. This type of biotechnology has long existed and is rapidly developing, particularly in areas of new and novel products, but there exists fierce competition between industrial concerns to achieve market advances. Product examples include antibiotics, monoclonal antibodies and interferons.

(b) *Medium scale biotechnology* will compete with petroleum-based technology to produce current commodity chemicals and with agriculture to produce natural products including proteins and lipids.

(c) *Large scale biotechnology* will compete with petroleum and coal to supply the primary sources of organic compounds for fuels and high tonnage industrial products.

Although medium and large scale biotechnology currently have only a few truly economic successes there can be little doubt that within the next two decades there will be the establishment of large scale microbiological processes using vegetable materials as feedstock (Figure 2). The market already exists for these products and the stimulus will be to develop economic biological alternatives.

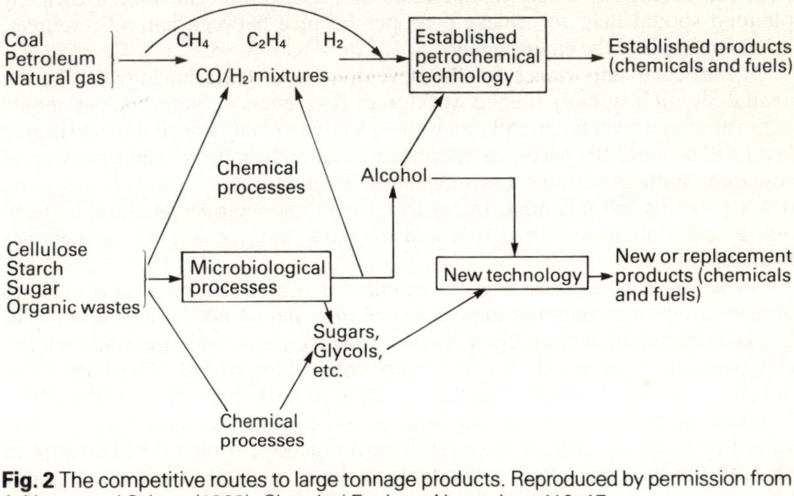

Fig. 2 The competitive routes to large tonnage products. Reproduced by permission from Atkinson and Sainter (1982), *Chemical Engineer* November: 410–17.

Growth in biotechnology

The continued successful growth of biotechnology will be dependent to a large extent on three premises, i.e. that there will be

(a) an expanding range of valuable products from both traditional and genetically engineered systems;
(b) the ability to obtain the raw materials from renewable resources;
(c) a growing awareness that biotechnological processes will in many cases become economical against the present chemical processing of vegetable materials.

A centrally important aspect of medium and large scale biotechnological developments will be the availability of suitable raw materials and substrates for processing. Substrate costs can amount to between 30 and 70% of the final product cost. Availability of substrates will be dependent on both technical and political factors. Thus gasohol production from various, organic substrates may not offer outstanding economic attraction but politically can present some bargaining power against escalating oil imports.

Renewable resources from agriculture, forestry and industrial organic wastes are gaining increasing importance since they offer, by biotechnological processing, a politically and strategically important base for food, feed, basic chemicals and energy production. However, to achieve this, wide ranging programmes not only in the component aspects of biotechnology but also in land management will be required.

Strategic planning in biotechnology

Each biotechnological scenario will require continuous assessment of the resources available, the economics and the impact on the environment, together with the health and safety of operators and users. Biotechnology if correctly planned should help to achieve a proper balance between natural resources, human needs and the environment.

Of particular importance for the development of biotechnology will be the availability of a suitably trained workforce. Awareness of biotechnology should begin at school level for all children while selective training at technical and degree level will produce the necessary specialists required by industry and by centres of research. Without suitable and sufficiently trained personnel biotechnology will not achieve its full potential. In particular high and intermediate level biotechnology can only proceed where a well trained workforce and management are available (Table 5).

Furthermore, when considering the application of biological discoveries to future industrial developments the importance of *time* should not be underestimated. Between the moment when a new discovery has been made and successful commercial production achieved, 5 to 20 years could have passed. Thus significant developments involving 'new' biotechnology must not be expected until the 1990s.

This short presentation will not attempt to describe or analyse the many and varied processes of current biotechnological practices. Rather it will attempt to identify the basic principles around which the core elements of biotechnology have developed.

Summary

Biotechnology can be considered to be the application of biological organisms and processes to manufacturing industries. Biotechnology encompasses a wide range of disciplines and subjects. Although present day activities are highly sophisticated and novel many of the processes have their roots in the dawn of history.

The specific processes are catalysed by microorganisms, plant or animal cells, or products derived from them such as enzymes. The organisms of biotechnology can be harvested for biomass, can be used to perform chemical conversions and may be the source of biologically active molecules, including enzymes and monoclonal antibodies.

Gene manipulation techniques have brought a new dimension to applied genetics and have created the potential for completely novel industrial processes, for example human interferon produced by bacterial cells. Significant developments are also occurring in process and control engineering and fermentation technology which will further advance the development of biologically-based industrial activity.

Biotechnology appears to be an area of expansion and opportunity involving many sectors of industry, including agriculture, food and feedstuffs, pharmaceutical, energy and water industries. It will play a major role in the production of new drugs, hormones, vaccines and antibiotics, cheaper and more reliable supplies of energy and (in the longer term) chemical feedstuffs, improved environmental control and waste management. Biotechnology will be largely based on renewable and recyclable materials thus being better fitted to the needs of a world where energy will become increasingly more expensive and in short supply.

References

ANON. (1982) Biotechnology market predicted to grow to $64B by 2000. *European Chemical News* 15 March.

ANON. (1984). *The FAST Report. Eurofutures – the challenges of innovation.* Butterworths, London.

ANON. (1984). *Splicing Life: The Social and Ethical Issues of Genetic Engineering with Human Beings*. President's commission for the Study of Ethical Problems in Medicine and Biomedical & Behavioural Research. Washington DC.

ATKINSON, B. and SAINTER, P. (1982). Downstream biological process engineering. *The Chemical Engineer* November: 410–9.

BULL, A. T., HOLT, G. and LILLY, M. D. (1982). *Biotechnology – International Trends and Perspectives*. OECD Publications Office, 75775 Paris, Cedex 16, France.

DEMAIN, A. L. (1983). New applications of microbial products. *Science* 219, 709–14.

DUNNILL, P. (1981), Biotechnology and industry. *Chemistry and Industry* April: 204.

DUNNILL, P, and RUDD, M. (1984). *Biotechnology & British Industry*. A report to the Biotechnology Directorate, SERC, London.

KIRK, T. K., JEFFRIES, T. W. and LEATHAM, G. F. (1983). Biotechnology: applications and implications for the pulp and paper industry. *Tappi Journal* 66, 45–51.

LYNCH, J. (1984). *Soil Biotechnology*. Blackwell Scientific.

PRAMER, D. (1983). Ensuring quality education in biotechnology. *Biotechnology* April 211–2.

REHM, H-J. (1980). *Industrielle Mikrobiologie*, 2nd edn. Springer, Berlin and Heidelberg, New York.

REHM, H-J. and REED, G. (1982). *Biotechnology*, Vol. I. *Microbial Fundamentals*. Verlag Chemie, Basel.

ROYAL SOCIETY REPORT (1981). *Biotechnology and Education*.

Scientific American (1981). Special Edition: Industrial Microbiology.

SHERWOOD, M. (1984). The case of the money-hungry microbe. *Biotechnology* July, 606–9.

SMITH, J. E. (1981). *Biotechnology*. Studies in Biology No. 136, Edward Arnold, London.

2 Applied genetics

The initial step in developing a biotechnological process is generally a search for a suitable organism. Such organisms will be expected to create a product or service that will generate a financial return to that industry. In practice, the geneticist has to select an organism that produces the desired product. Once a suitable organism is found, conventional breeding and mutagenesis methods are used where possible to induce genetic changes that may produce even more of the desired product. Selection of improved organisms is tedious and time demanding and most of the methods available to the geneticists up until recently have involved trial and error. However, new genetic technologies, i.e. protoplast fusion and the use of recombinant DNA techniques, are allowing new approaches by which useful genetic traits can be inserted directly into the chosen organism. Thus, totally new capabilities can be engineered and microorganisms in particular and plant and animal cells to a lesser extent may be made to produce substances beyond their naturally endowed genetic capabilities.

Increased productivity is not the only goal of the applied geneticist when manipulating a potential industrial organism. Thus, resistance to viral infection and increased genetic stability may be incorporated into organisms that lack them, the formation of harmful by-products can be reduced or eliminated and objectionable odours, colours or slime products can be removed.

A successful industrial culture should ultimately exhibit most or all of the following characteristics: it should be a pure culture; be genetically stable; be easily propagated; exhibit rapid growth characteristics; have good rate of product formation; be free of toxic byproducts; and be amenable to genetic manipulation.

Cultures that are used industrially generally have arisen in three stages: (1) as a research culture—studied to seek a useful product: (2) as a development culture—a research culture which has gained a measure of importance; and (3) as a production culture—a research culture that is now actually used for industrial production.

The final culture may be the same as the original research culture but more often it will have been subjected to an array of treatments to maximize productivity. This final industrial organism will have been genetically programmed to perform a metabolic function far in excess of that of the wild type. This will only be achieved by maximum control of the growth of the organism during production. The development of the industrial organism from a wild type will have required changes in its genetic information that eliminate undesirable properties or even introduce entirely new ones.

Finding the desired organism and improving its capabilities is now a fundamental aspect of most biotechnological processes (Figure 3).

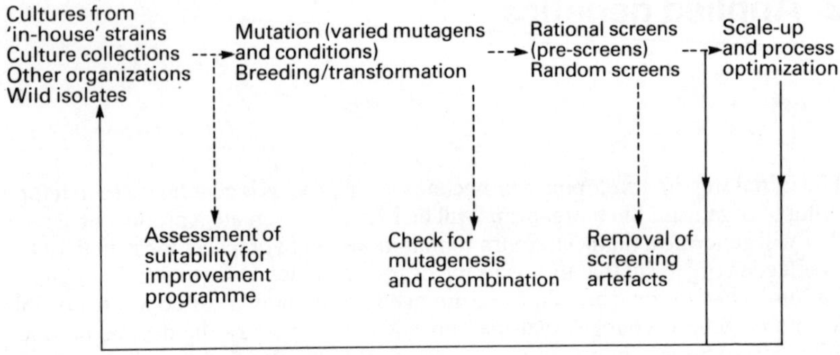

Fig. 3 Flow diagram for strain improvement. Reproduced by permission from Rowlands (1983).

Selection and screening

In all aspects of biotechnology major efforts are directed to screening programmes to generate new organisms, either from some natural source or from established cultures by way of mutation or by hybridization programmes, including genetic engineering. These organisms must be screened for useful products and grown up on a large enough scale to produce and extract the desired product and then to subject the product to critical evaluation. Screening can be defined as the use of highly selective procedures to allow the detection and isolation of only those microorganisms or metabolites of interest from a large population. The further advancement of an isolate will involve improvement and preservation of the culture. However, a major hindrance to the full exploitation of this capacity is the availability of suitable screening procedures which can identify the necessary product, especially in the presence of culture medium constituents.

The major group of organisms presently used in biotechnology are microorganisms. The screening methodologies to be described will concentrate largely on this group.

In the search for new microorganisms from the environment for biotechnological processes there are normally three types of option available; these involve the choice of habitat for sampling, the physical separation procedures for separating out the desired microorganism and the choice of method to achieve selection which in most cases involved enrichment cultures (Table 6).

Although many new producer microorganisms are wild-types and have been isolated from natural environments, major efforts are also directed to generating new genomes from existing genomes by laboratory manipulation. Organisms can be modified by mutation, recombination, transformation, transduction and gene cloning either by single processes or in combinations (Table 6).

With natural selection and, more particularly, with genetic manipulation, all industrially important microorganisms will have been subjected to some form of screening. The design of the screening programme is of major importance to achieve maximum recognition of new genotypes. Screens can be divided into two basic forms:

Table 6 Generation of producer microorganisms

Ecological approach	Gentic approach
Strategy	*Strategy*
Set up isolate library of known species	Generate biosynthetic mutants suitable for precursor feeding
Enrich isolate library in rare species	Prepare hybrid strains to produce antibiotics
Explore unusual habitats and develop isolation techniques for new species	Use gene cloning to redirect biosyntheses
Dispersal Techniques	*Techniques*
Soil dilution	Mutation and selection
Flotation	Recombination by
Particle sedimentation	Conjugation
and air sampling	Protoplast fusion
Microhabitat outgrowth	Fungal, sexual and parasexual cycles
Selection and enrichment techniques	Transformation and transduction
Selective substrates, e.g. chitin	(enhance by using protoplasts)
	Gene cloning by
Selective inhibitors, e.g. cycloheximide, rifampicin	Target-directed cloning of specific genes
Physical environment, e.g. temperature, pH, Eh	Random shotgun cloning to generate a gene library

Adapted from Nisbet, (1982).

(1) *non-selective random screens* with all the isolates being tested individually for the desired qualities; and
(2) *rational screens* in which there is some aspect of preselection.

Random screening can be very time-consuming because each isolate must be studied in detail. In studies of antibiotic production thousands of shake-flask cultures of single spore isolates are regularly used. Agar disc techniques have accelerated the process but this method is best used as a primary screen before using the shake flasks. Large numbers of isolates or mutants can now be plated out, exposed to a wide range of environmental conditions and the responses recorded automatically over any time period using television monitoring and computer control.

In contrast, rational screens make more use of the knowledge of the biochemistry of specific product formation, setting up a selection process that uses a characteristic of the desirable genotype which is not the particular one of interest but which can be more readily scored or assayed. At its best the rational screen should selectively kill all undesirable genotypes thus permitting much higher numbers of isolates to be tested routinely. Thus antibiotic production in *Cephalosporium acremonium* has recently been shown to correlate with proteolytic activity suggesting a possible basis for a screen.

After screening procedures have been employed and potentially worthwhile microorganisms isolated the final proof of efficiency will come under production conditions. Product optimization from new and existing microorganisms also involves selection of the best method of cultivation, solid or liquid, batch or continuous. The type of medium used to propagate the organism can have a major influence on phenotypic expression of product formation. Thus medium composition can influence biomass concentration, specific rate of product formation, duration of product formation, rate of decomposition of the product and stability of the producer organism. However, the production environment cannot remain static and successful production may well also require modification to the medium and the environmental control parameters.

Strain degeneration is a regular problem encountered in biotechnology. Stability is genetically controlled and can be modified by specific control of environmental factors such as medium or by the inhibition of the genetic events that lead to instability.

Culture maintenance

Having generated a novel microorganism either by natural selection or by genetic manipulation, the new organism must be stored or preserved with minimum degeneration of its genetic capabilities. Maintenance, preparation and propagation of the organism must achieve specific standards of reproducibility. The maintenance process of an industrial microorganism is an intergral feature of the infrastructure of biotechnology. There are no methods common to all industries. Specific maintenance techniques for industrial microorganisms are usually well kept industrial secrets. In practice most industrial microorganisms will be preserved by any one of the following procedures:

(1) *on agar medium* with regular subculturing;
(2) *by reduced metabolism*—mineral oil coverage, refrigeration, or storage in a deep freeze;
(3) *drying*—dry sand, silica gel, soil or filter paper;
(4) *freeze drying*—widely practised because of convenience and gives greater stability than previous methods;
(5) *cryopreservation at ultra low temperatures* (−70°C to −196°C—a more costly method but suitable for a wide range of organisms and one that gives high survival rates).

Culture collections throughout the world are playing an increasingly important role in biotechnology in providing a comprehensive range of pure, verified organisms that are of past, present or potential interest. Important operating strains must be retained in a viable and productive condition. In particular, many new genetically manipulated organisms exhibit some degree of instability and preservation must aim at achieving minimal strain drift.

The remaining part of this chapter will examine in some detail the various techniques that are now available to the applied geneticist for modifying the genome of industrially important organisms.

Mutagenesis

Once a useful organism has been isolated by preliminary screening several options are available for improving productivity, which include modifications to the medium and fermentation conditions (Chapter 3), mutation (spontaneous or induced) and hybridization. Since mutation is the ultimate source of all genetic variation it has become a topic of fundamental importance in industrial microbiology. Furthermore, many of the important microorganisms used industrially (*Penicillium chrysogenum, Aspergillus oryzae*) do not exhibit normal sexuality and therefore do not lend themselves easily to hybridization. For this reason there has been considerable use of various mutagenesis techniques to produce many of the currently used production strains. Indeed, for many industrial processes, mutagenesis programmes represent the main approach to increased productivity. Mutagen-induced improvements in productivity or titre have long been practised for antibiotic production in *Penicillium, Cephalosporium* and *Streptomyces*, for organic acid production and enzyme production with *Aspergillus* species, amino acid production by various bacterial strains and for numerous other industrial microorganisms. Mutagenesis cannot yet be routinely supplanted by the newer techniques of genetic engineering.

It is now recognized that the most effective method for increasing the yield of a fermentation product is by use of induced mutations followed by selection of the improved strains. The main difficulty is that mutations occur at low frequencies and have to be selected from large non-mutant populations.

There are many examples of major mutational changes leading to improved industrial strains producing more efficacious products. Tetracycline producing strains of *Streptomyces* have been particularly significant in this area. A mutant strain of *S. aureofaciens* S-604 was found to synthesize 6-demethyltetracycline which was not synthesized by the parent production strain. This mutant is now one of the main commercial producers.

However, minor mutations have normally been the main approach to strain improvement. Such mutations normally lead to small improvements in product formation (5–10%) without any phenotypic manifestations. Successive use of small mutational changes can lead to an increase in the concentration of the hereditary factors responsible for the productivity of the original genotype. The cephalosporin-producing organisms have been of special interest here together with the penicillin producing *Penicillium chrysogenum* strains.

Selection of mutagens Although mutagens can be classified into physical (radiation) and chemical sources, this is not of much relevance to the industrial geneticist. The main reason for choice will be the ability of the chosen technique to generate as wide a spectrum of mutants as possible for purposes of selection. Many induced mutations are not directly the result of the type of damage caused to the DNA but rather the result of the cellular DNA repair processes acting on the damage to make it into a fixed alteration in the base sequence. Mutagens which induce this phenomenon include UV, ionizing radiation, thymine starvation and certain chemical mutagens such as mitomycin C, 5-bromouracil, methylmethane sulphonate, nitrogen mustard and the nitrofurans.

Knowledge about the molecular basis of mutagen specificity is incomplete, but it is substantially dependent on mutagenic repair processes. In practice, producer strains can become refractory to a particular mutagen and it is advisable to alternate between several mutagens which function through different repair pathways. Furthermore, correct choice of mutagen dose can considerably increase the frequency of the desired type of mutant in the mutagenized population and in this way increase the chances of identification when the whole population is screened.

Mutation rate in an organism is genetically controlled and can be changed by mutator or antimutator genes. These genes are able to affect spontaneous or induced mutability or both. In screening programmes mutator strains are important since some will give a high frequency of mutants per survivor after mutagenesis and this will provide a most effective input.

The environment immediately after mutagenesis can affect the overall mutation frequency and specificity. Thus growth on a complete medium rather than a minimal medium following mutagenesis may increase the yield of mutants.

Most living organisms have a mechanism for exchanging nuclear material with other similar but genetically different individuals resulting in the formation of offspring with genotypes different from either of the parents. This process is known as hybridization and is essentially a method of promoting recombination of the available genetic material. Whereas mutation alters an organism's genes, recombination or hybridization rearranges genes or parts of genes and brings together in an individual organism genetic information from two or more organisms. Hybridization has long been practised in plant and animal breeding and in more recent times has been used to develop improved microorganisms for many biotechnological processes. In principle, hybridization can be expressed in two ways; these are sexual hybridization with eukaryotic organisms and parasexual hybridization for prokaryotes, certain eukaryotes that do not exhibit true sexuality and for most eukaryotic tissue cultures. In the context of biotechnology, fungi and bacteria have been the main subjects of hybridization programmes.

Sexual hydridization

In sexual hybridization haploid nuclei from opposite mating types are brought together in one cell (karyogamy): the nuclei will ultimately fuse to form the diploid nucleus and will then undergo meiosis. During the meiotic divisions there will be a rearrangement and reorganisation of the chromosomes resulting in recombination of the genetic elements (Fig. 4). This homologous recombination is very effective in producing new genotypes. Thus if two organisms differ in n genes, recombination between their sets of genes will generate 2^n genotypes. If the two strains differ only in a dozen base pairs scattered among many millions of base pairs this would generate 2^{12} or 4096 new genotypes.

Sexual hybridization can be complicated in many organisms by the existence of breeding systems which regulate the outbreeding of a population by impeding inbreeding through self-fertilization. In this way breeding systems encourage a continuous mixing of suitable genetic material. Breeding systems have been well-documented for many higher Ascomycetes and Basidiomycetes, and have been exploited extensively in the commercial production of various mushrooms: *Agaricus bisporus, Lentinus edodes* and *Volvariella volvacea*.

Sexual hybridization of different strains of the yeast *Saccharomyces* has been used for the rapid production of bread by modern factory methods, for increasing the alcohol content of liquors for distillation and for brewing special beers from which nearly all the soluble carbohydrates have been removed.

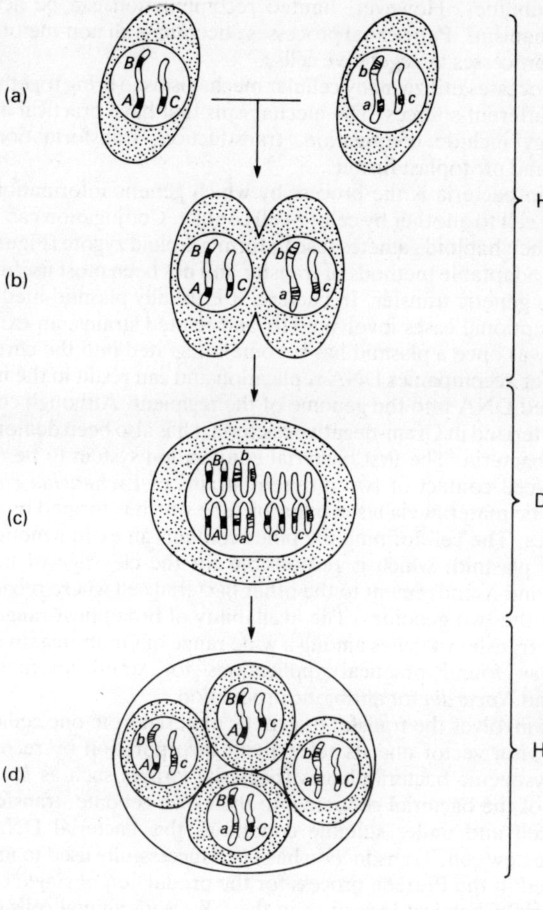

Fig. 4 Meiosis in eukaryotes. Recombination in eukaryotes involves crossing-over between parts of chromosomes and the reassortment of chromosomes. A typical yeast is haploid during most of its life cycle, with a single set of 15 or more chromosomes; only two are shown here. Two cells of opposite mating types can fuse (a), then then nuclei fuse (b) to form a diploid nucleus with two complete sets of chromosomes. During meiosis (a phase of sexual reproduction) the chromosomes become double structures consisting of two chromatids (c). Homologous chromosomes pair and exchange parts of their chromatids by crossing-over. Then four haploid sexual spores are formed (d). Each spore can have a new combination of genes that were different (black and barred) in parent cells; genes on the same chromosome (A, a; B, b) recombine by crossing-over and genes on different chromosomes are shuffled as members of chromosome pairs reassort. (From Hopwood, 1981.)

Parasexual processes

Few of the important organisms used in current biotechnology exhibit overt sexual recombination abilities. However, limited recombination can be achieved by parasexual mechanisms. Parasexual processes encompass all non-meiotic, genetic recombination processes in vegetative cells.

Parasexual processes utilize many cellular mechanisms to bring together genetic material from different sources. The mechanisms that have practical application to biotechnology include conjugation, transduction, transformation, mitotic recombination and protoplast fusion.

Conjugation in bacteria is the process by which genetic information is transferred from one cell to another by cell-to-cell contact. Conjugation can also occur in eukaryotes when haploid gametes fuse to form a diploid zygote (Figure 5a). It is one of the most adaptable methods of transfer and has been most useful for intra- and interspecies genetic transfer. In bacteria it is usually plasmid-mediated and, even in the exceptional cases involving Hfr and related strains, an extra genetic element which was once a plasmid has become integrated into the chromosome. Conjugal transfer accompanies DNA replication and can result in the integration of the transferred DNA into the genome of the recipient. Although conjugation has been characterized in Gram-negative bacteria it has also been demonstrated in Gram-positive bacteria. The first bacterial conjugation system to be discovered involves cell-to-cell contact of two different strains of *Escherichia coli* and the transfer of genetic material via a sex pillus or tube which is formed by one of the two mating types. The cell forming the pillus contains an extra genetic element, the *F*-factor or plasmid, which is responsible for the cleavage of the circular bacterial DNA and its movement to the other bacterial cell where recombination occurs between the two genomes. The availability of broad host-range plasmids has allowed the transfer of genes among a wide range of Gram-negative bacteria. Conjugation has found practical applications for strain improvements of *Streptomyces* and *Nocardia* for antibiotic production.

Transduction involves the transfer of genetic material from one cell to another by means of a viral vector and its subsequent incorporation by recombination (Figure 5b). Lysogenic bacteriophages (bacterial viruses such as *lambda*) may integrate parts of the bacterial genome into their own genome, transfer this into another host cell and under suitable conditions this bacterial DNA may be expressed in the new cell. Transduction has been successfully used to improve the bacteria involved in the Pruteen process for the production of single cell protein (SCP) by Imperial Chemical Industries in the UK. With animal cells the *Simian*

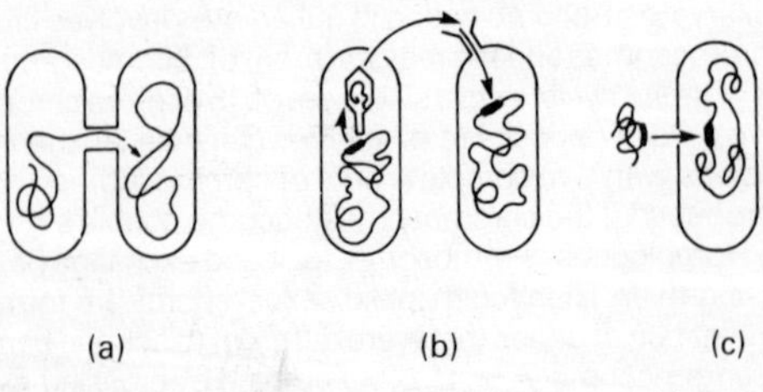

Fig. 5 Mechanisms of parasexual processes: (a) conjugation, (b) transduction, (c) transformation.

virus 40 (*SV40*) and the mouse virus polyoma are gaining in interest while for plant cells one type of viral transfer agent has been of particular interest. This is the cauliflower mosaic virus (CMV), a double-stranded DNA molecule which is restricted in host range to the Cruciferae. It is not anticipated that this viral system will have major use in plant biotechnology.

Transformation, or the unidirectional transfer of genetic material in which DNA originating from one cell is taken up and stably maintained by another, has been widely used with bacteria (Figure 5c). It is being developed extensively and rapidly with yeast and *Streptomyces* but so far with only limited success in filamentous fungi and plant and animal cells. The transforming DNA can be subjected to *in vitro* treatments or manipulation including mutagenesis before introduction into cells. The main disadvantage of transformation is that for most bacteria only a small proportion of cells in a population will be transformed (about 0.1% with *E. coli*, 10 to 100 times higher with *B. subtilis*). It has been shown with *E. coli* that large plasmid DNA molecules (>20 kb) transform relatively inefficiently compared to smaller molecules. Transformation has become the main method of genetic engineering using plasmids as transport vehicles and is discussed later in this chapter.

Mitotic recombination has been of particular interest with filamentous fungi. With certain fungi where the vegetative phase is haploid small numbers of the nuclei can fuse to form diploids. This diploid phase in the asexual cycle of the fungus results from the fusion of vegetative nuclei which are in the heterozygous condition. The process consists of the following stages: heterokaryon (two or more different nuclei with a single cytoplasm) formation between two haploid mycelia; nuclear fusion by a small number of the opposite nuclei (frequency of 10^{-6}); mitotic crossing over within the diploid nucleus (frequency about 10^{-2} per nuclear division) giving interchromosomal exchange; and haploidization where the diploid nuclei are reduced to the haploid condition at a fequency of about 10^{-3} per nuclear division (Figure 6). This mitotic recombination makes possible genetic analysis and controlled breeding in organisms with no sexual cycle. With respect to efficiency of recombination mitotic recombination is very much less effective than sexual recombination. This form of parasexual hybridization is known to occur in many industrial fungi including *Penicillium chrysogenum, Aspergillus oryzae* and *Cephalosporium acremonium*. It is difficult to know how widely this system has been used in industrial practice due to the requirements of industrial secrecy. However, there is little doubt that it has been successfully used with citric acid and penicillin producing organisms. Mitotic recombinations have also been recorded for some higher plants and animal tissue cultures.

The primary necessity for achieving parasexual hybridization is the fusion of the desired cells. Many barriers exist due to homogenic and heterogenic incompatibility. However, these incompatibility or species barriers can now be overridden in many cases by specific laboratory procedures which permit protoplasts from cells to fuse directly or to be infected with isolated DNA by way of transformation.

Protoplast fusion. The main natural barrier to recombination between dissimilar organisms is usually the cell wall. If this barrier is removed by preparing protoplasts and recognizing that cell membranes have relatively similar compositions, it is feasible that protoplasts of different species could be induced to fuse and form a hybrid cell, in this way allowing their genes to recombine. The field of induced

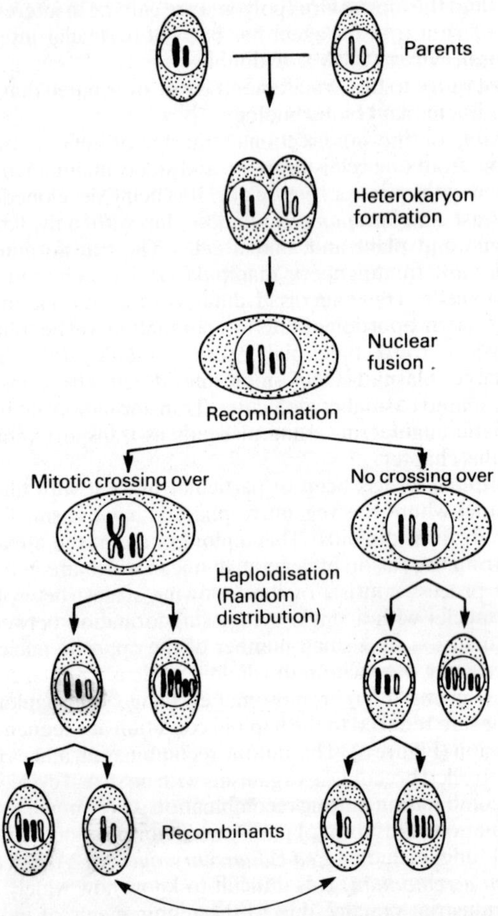

Fig. 6 Scheme of mitotic recombination. Reproduced by permission from Esser and Strahl (1981).

fusion of microbial protoplasts is one of great expansion and there has been a rapid accumulation of data in both basic and applied areas. The present success in protoplast fusion and regeneration technology with bacteria, fungi and plant and animal cells owes much to the many earlier studies establishing the basic techniques of protoplast release and cell regeneration. Protoplasts can now be routinely formed from most microbial cells.

Although mechanical and other non-enzymic methods for protoplast release have been reported, the main directions of study have favoured use of lytic enzymes for protoplast isolation. The principal components of this lytic digestion are the enzymes, the osmotic stabilizer and the organism.

Yeast protoplasts can be produced using commercial preparations of snail digestive juice which is available as helicase, glusulase or sulphatase. In general, most commercial lytic preparations have not been very effective for preparation of protoplasts from filamentous fungi. However, many other organisms, in particular *Streptomyces* spp. and certain filamentous fungi, can produce highly effective mycolytic activity. In most cases the synthesis of the wall digesting enzyme complex by the producer organism requires the presence of inducer substrates in the growth medium. A novel approach to the production of wall digesting enzymes is to grow the producer organism in a medium low in normal carbohydrate source such as glucose or sucrose but containing relatively purified cell walls of the subject organism. Successful growth of the producer organism will require appropriate production of wall digesting enzymes in order to generate essential low molecular weight energy sources.

In general, exponential phase cells produce protoplasts more readily than stationary phase cells, suggesting that wall chemistry changes during the growth cycle.

Chemical stabilizers are also essential to provide osmotic support to the protoplasts after the removal of the cell wall. Inorganic salts, sugars and sugar alcohols have been successfully used for retaining the integrity of the released protoplasts. Without osmotic stabilizers, released protoplasts will rapidly burst.

A high proportion of protoplasts can generate a new cell wall and subsequently a full reversion to a normal growth can be achieved. Thus if mitotic recombination can be achieved between isolated fused protoplasts, the ensuing reversion could lead to a genetically altered and viable organism.

Fusion between natural protoplasts of similar or dissimilar strains occurs at a very low frequency. However, with the introduction of polyethylene glycol (PEG) as a fusogenic agent, the frequency of fusion has been enhanced at least 1000-fold. Methodological studies on PEG-induced fungal protoplast fusion has shown that PEG preparations with molecular weights of 4000 or 6000 were optimal in fusion induction in the concentration range 25 to 40%. The presence of Ca^{2+} ions is critical for the attainment of high-frequency fusion. In practice fusion events start with agglutination of the protoplasts caused by extensive dehydration and formation of various aggregates. Protoplasts then shrink and become highly distorted. Translocation of intramembrane protein particles occurs at the sites of close contact, and is followed by their aggregation. Lipid–lipid interaction occurs between adjacent protein-denuded membranes and there is certain reorganization of the lipid molecules assisted by calcium ions. This results in fusion of small areas of membranes in contact: small cytoplasmic bridges are formed which enlarge until finally the two protoplasts fuse.

A novel electrofusion method for protoplast fusion has recently been developed. Electric field fusion is a rapid and gentle process used to fuse like cells into giant cells, unrelated cells into hybrids and to encapsulate substances within cells. Cells are exposed to low-level, non-homogenous, high frequency electric fields which orientate the cells into chains. A direct current is then applied which opens micropores in adjoining cell membranes, allowing intermixing of the cellular contents and cell fusion. This process can also alter the permeability of the outer cell membrane to permit substances the size of genes to be encapsulated.

When two protoplasts derived from different strains fuse, the resultant cell is heterokaryotic since it will contain nuclei from different genetic sources. In the heterokaryon, nuclear fusion or diploidization can take place between haploid nuclei of different genotypes as well as between nuclei of the same type. The former results in a heterozygous diploid nucleus while in the latter a homozygous diploid nucleus will be formed. The fused protoplasts with their mixed complement of haploid and diploid nuclei can then be made to regenerate into a growing organism with normal means of propagation.

During this growing out phase the diploid nuclei may undergo division with occasional mitotic crossing-over. As a result new combinations and linkages will occur and these recombinations give the organism some of the advantages of normal sexuality.

By way of this parasexual process new diploid strains of previously haploid organisms can be formed. However, more often haploidization of the diploid colonies will occur. Some of the new haploid strains will be genotypically different from either parent because of mitotic recombinations producing new linkage groups. In filamentous fungi haploidization more often shows up in conidia formation where individual conidia are uninucleate and haploid.

When protoplast fusion is to be used to improve industrial strains three events must occur:

(a) the generation of viable protoplasts,
(b) high fusion frequency,
(c) regeneration of the fused protoplasts to growing colonies in which diploidization and haploidization can be realized.

Fusion of protoplasts seems to be the most promising method for strain improvements of *Streptomyces* and other known producers of antibiotics. This could well lead to the generation of new antibiotic structures and to the increase in the pool of yield-enhancing genes. Recently two different types of yeast have been fused and the regenerated cells propagated for use in sherry production. For this system cell fusion techniques are more promising than recombinant DNA techniques since there are large numbers of genes involved in such improvements. Protoplast fusion in filamentous fungi together with inter-specific protoplast fusions has produced successful strain improvements. Protoplast fusion seems to be the only means of obtaining interspecific heterokaryons and recombinants in filamentous fungi that do not have a demonstrable sexual cycle.

Protoplasts can be prepared from most plant species but only a limited number can regenerate cell walls, divide and ultimately differentiate into normal plants. Many important crop plants are in the latter category. The production of novel somatic hybrids by fusion of isolated plant protoplasts is recognized as an important technological advancement in the plant sciences.

In practice, following induced fusion, protoplasts in liquid culture regenerate cell walls and undergo mitosis, creating a mixed population consisting of parental cells, homokaryotic fusion products and heterokaryotic fusion products or somatic hybrids. Somatic hybrids can be identified and recovered generally through expression of genetic markers, physiological growth requirements or alternatively by micromanipulation. Protoplast fusion between divergent species can sometimes produce hybrids but these are invariably sterile, precluding integration into a breeding programme. A role for somatic hybrids in crop improvement can only be realized if the hybrids can be integrated into conventional breeding programmes. As yet no new varieties have been established in agricultural practice by protoplast fusion technology. However, this will undoubtedly occur in the near future and will open up a new dimension of creative and beneficial agricultural advancement.

Protoplast fusion is also possible with animal cells but there is no opportunity to have complete organism development. However, both with plant and animal tissue propagation, protoplast fusion can be a valuable means of changing the genetic complement.

Unusual examples of intergenetic and inter-kingdom protoplast fusions have occurred, e.g. yeast protoplasts with hen erythrocytes, animal–fungal heterokaryons, and bacteria with yeast. However, there have so far been no immediately worthwhile genetic consequence resulting from these experiments, but they do suggest that the future may offer some potential for fusing taxonomically quite distant cells with the possibility of specific transfer of genetic elements.

Cell fusion Direct cell fusion has formed the basis of monoclonal antibody formation—one of the most active and successful areas of new biotechnology. Monoclonal antibodies are antibodies derived from a single source or clone of cells which recognize only one kind of antigen.

A longstanding area of health care has been the preparation of vaccines to combat disease. Vaccines allow the body to build up resistance, i.e. to build up an adequate level of antibody and a primed population of cells that will grow and produce antibodies when the pathogen occurs in a virulent form. Vaccines are complex entities made up of a mixture of antibodies too complex to be synthesized chemically.

When a foreign substance or microorganism (antigen) is introduced into a vertebrate animal the immune response reacts by forming plasma cells (derived from B-lymphocytes) which secrete antibodies that recognize, neutralize and eliminate the antigen with great specificity. In practice the antibody response to a particular antigen is highly heterogeneous. It is practically impossible to separate these antibodies and so conventional antisera contain mixtures of antibodies. However, each antibody is now recognized to be made from a different line of B-lymphocytes and their derived plasma cells. Thus if one cell line could be cloned it would be possible to produce only one antibody. Unfortunately, plasma cells cannot be cultured. In 1975 Kohler and Milstein in their now famous experiment demonstrated that antibody producing lymphocytes could be fused with malignant, rapidly proliferating myeloma cells and the hybrid-myeloma cells or hybridomas could express both the lymphocyte-specific antibody production and the myeloma property of continuous proliferation. Thus was born the monoclonal antibody.

Monoclonal antibody formation is carried out by injecting a mouse or rabbit with the antigen, removing the spleen and then allowing fusion of spleen cells with myeloma cells (Figure 7). About 1% of the spleen cells are antibody secreting plasma cells and 10% of the final hybridomas consist of antibody secreting cells. Each hybridoma cell is capable of producing only one antibody and selection techniques can be developed to identify them and then to carry out further propagation or cloning. Clones derived from monoclonal antibody hybridomas may be kept frozen, injected into an animal or grown in culture where the clone produces the antibody in large amounts. Many of these clones are genetically unstable and lose chromosomes to become more stable, but more often the chromosomes lost are those which encode for the desired antibody.

Monoclonal antibodies have a wide range of uses as very precise and sensitive analytical reagents for various pharmaceutical products such as peptides, proteins and hormones. They are also important as diagnostic reagents for a number of medical and therapeutic purposes, including screening for the presence of

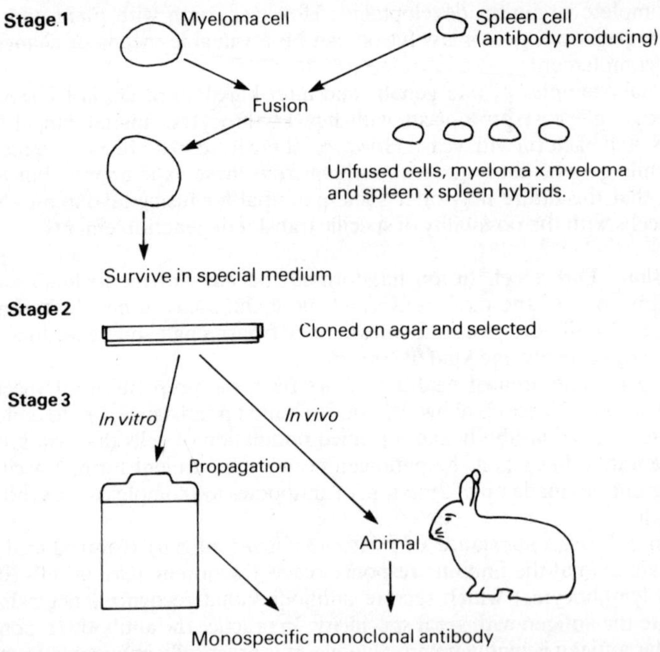

Fig. 7 Diagram illustrating the formation of antibody-producing hybridomas by fusion techniques.

undesirable agents in blood samples used for transfusions. A combination of genetic manipulation and monoclonal antibody techniques might well give rise to new approaches to efficient production of viral, bacterial and parasitic antibodies. Other important uses will include tissue typing, *in vivo* tumour location, clinical diagnosis and therapy including targeting of chemotherapeutic agents.

There is now little doubt that fusion technologies will contribute substantially to the biotechnological utilization not only of prokaryotes and simple eukaryotes like yeast and filamentous fungi but also for higher plants and animal tissue cultures where specific products are being sought.

Recombinant DNA technology

The analysis and manipulation of genetic material has been revolutionized by the development of *in vitro* gene cloning techniques which have allowed the isolation, purification and selective amplification in suitable biological systems of almost any discrete segment of DNA from almost any organism. Due to deletion of DNA not containing relevant genes, gene cloning enables the methods employed for analysis and manipulation to be targeted primarily on the gene region of interest, thus vastly increasing the efficiency of the method used and simplifying the identification and characterization of the now modified derivatives. In contrast to protoplast fusion where large parts or entire genomes are mixed and the characteristics of interest are controlled generally by a large number of genes (for example, antibiotics), recombinant DNA technology is primarily concerned with small numbers of individual genes controlling known gene products (for example, proteins, peptides, etc.).

The 1978 Genetic Manipulation Regulations defined genetic manipulation as 'the formation of new combinations of heritable material by the isolation of nucleic acid molecules, produced by whatever means outside the cell, into any virus, bacterial plasmid, or other vector system so as to allow their incorporation into a host organism in which they do not naturally occur but in which they are capable of continued propagation'. Genetic manipulation thus defined does not include, for example, monoclonal antibody work, strain improvement of microorganisms by conventional genetic techniques, and *in vitro* fertilization or cloning techniques for animal use. Genetic manipulation techniques have so far made only a small contribution, although an important one, to biotechnological research and applications.

Over the last ten years several important discoveries were made in molecular biology which led to the rapid development of recombinant DNA technology. Of particular significance were the isolation of a mutant strain of *Escherichia coli* unable to break down or restrict foreign DNA introduced into the cell, and the identification and characterization of restriction endonucleases—enzymes which are able to cleave double-stranded (ds) DNA at specific sites. In general, such DNA fragments are unable to transform or replicate efficiently in host cells. This problem has been largely overcome by inserting the DNA fragments into vectors which can be inserted into host cells and which allow the relevant information to be expressed. The vectors are usually bacterial plasmids, phage or yeast 2μ DNA. Covalent bonding of the DNA fragments into the vector is achieved with the enzyme DNA ligase. The newly synthesized vectors or chimeric DNA molecules are introduced by transformation into host organisms such as *E. coli, Bacillus subtilis* or *Saccharomyces cerevisiae* and as the host cells grow and multiply (cloning) so also will the vectors replicate and amplify the foreign DNA. It is possible to clone an entire genome, thus creating a large collection or library of independent DNA molecules. A genomic library will contain all the sequences of the organism genome and in principle it should be possible to isolate any gene provided that a specific probe for detecting it can be obtained. If the average size of a cloned sequence is 20 kilobases, approximately 700 000 individual clones would be required for a complete library of the human genome.

Genetic manipulation techniques offer revolutionary abilities because they are rapid, universally applicable, allow a high degree of control over the manipulation process and above all generate the formation of new genetic combinations which have never before been selected by laboratory methods. Recombinant DNA technology has been one of the most rapidly developing aspects of bioscience.

Central to this new technology is the concept of cloning which has been defined by Timmis (1981) as 'the separation and individual propagation of a single element that is capable of reproduction from a mixture of similar elements'. In practice, cloning has long been used to propagate pure cultures of viruses, microorganisms and plants by vegetative means thus ensuring exact continuation of the organism's genome (Table 7). Nowadays, the term also encompasses the isolation and maintenance of individual autonomous genetic elements such as plasmids and cryptic virus genomes. Cloning of cellular organisms needs only a nutrient medium; cloning of viruses and plasmids will involve a specific host cell together with a

Table 7 Cloning

Origin of genes	Typical habitat	Separation	Growth medium	Amplification	Requirements
Bacteria	Soil	Physical, by dilution, on solid medium	Nutrient agar plate	Of individual bacteria to form colonies	Nutrient medium
Viruses	Sewage	Physical and biological, by dilution, infection of sensitive host cells by individual virus particles	Pure culture of host bacteria; nutrient agar plate	Of individual viruses to form plaques in growing lawn of host cells	Host bacteria; nutrient medium
Plasmids	Bacterial lysate	Physical and biological, by dilution, transformation of competent cells by individual molecules	Pure culture of host bacteria; nutrient agar plate containing agent selective for a plasmid-encoded property	Of individual plasmid molecules in individual bacteria that form colonies	Host bacteria; nutrient and selective medium
Genes*	Chromo-somes	Biochemical or physical; by breakage into gene-sized fragments and ligation into plasmid molecules and subsequent transformation of competent cells	Pure culture of host bacteria; nutrient agar plate containing agent selective for a plasmid-encoded property	Of individual plasmid molecules in individual bacteria that form colonies	Cloning vector; appropriate DNA fragmentation procedure; DNA ligase; host bacteria; nutrient and selective medium

*This entry refers to the *in vitro* cloning of genes. From Timmis (1982).

nutrient medium, while cloning of genes will require a vector replicon, a specific host cell and a nutrient medium. All types of cloning have important roles in biotechnology; thus all microorganisms, plant and animal cells to be used in biotechnological processes need to be maintained and propagated in a pure and constant state, while viruses and plasmid propagation or amplification, in particular, have become identified as powerful tools in the new biotechnology.

The principles of gene cloning or recombinant DNA technology will be briefly discussed using mainly the example of the *Escherichia coli* K-12 plasmid vector system. Although gene cloning techniques using viral vectors differ in detail, the principles involved are similar.

The basic requirements for *in vitro* transfer and expression of foreign DNA in a host bacterial cell are outlined in Figures 8 and 9 and can be summarized as follows:

(1) The isolation of circular plasmid DNA molecules which must contain one site where the integration of foreign DNA will not destroy essential functions.
(2) The generation of DNA fragments that are suitable for cloning by way of restriction endonucleases. The insert may be a chromosomal fragment from another microorganism (prokaryote or eukaryote), from an animal or plant, or from a chemically-sythesized DNA sequence.
(3) A method of splicing the foreign DNA into the vector.
(4) Incorporation of the hybrid DNA recombinants into the host cell.
(5) Expression of the transformants which have the recombinant plasmids.

Thus a major feature of this new genetic technique is that it allows considerable control over the DNA segment to be cloned and can be used essentially for any DNA fragment from any organism.

1 Plasmids Fundamental to this new technology are plasmids. Plasmids are basically molecules which can be stably inherited without being linked to the chromosome. They are important in medicine and agriculture since they confer antibiotic resistance to pathogens of man and animals and can code for toxins and other proteins which may increase the virulence of such pathogens. They are also important since some may be involved in nitrogen fixation in *Rhizobium sp*. while others can code for a wide range of metabolic activities that allow certain bacteria to degrade compounds that could become environmental pollutants. Plasmids are molecules of double-stranded DNA which exist as covalently closed circular molecules (CCC) in their host cells. The molecular weight of plasmids ranges from about 1×10^6 to 200×10^6. Conjugative plasmids can transfer copies of themselves from one bacterium to another and can pass pieces of chromosomal DNA between bacteria. Insertion sequences and transposons can also be carried on plasmids. Plasmids can be used as vectors to clone DNA.

Plasmids carrying antibiotic resistance genes are termed resistance (R) plasmids and certain of the *E. coli* R plasmids have been used as plasmid vectors. One of the most frequently used plasmids for cloning *E. coli* genes is pBR322 which confers resistance to ampicillin and tetracycline. Plasmid DNA can be isolated by lysing the bacterial cells and separating plasmid DNA from chromosomal DNA in a caesium chloride gradient containing ethidium bromide.

Some of the smaller plasmids such as Col E1 are maintained at high copy numbers in host cells and can therefore synthesize large amounts of plasmid-coded

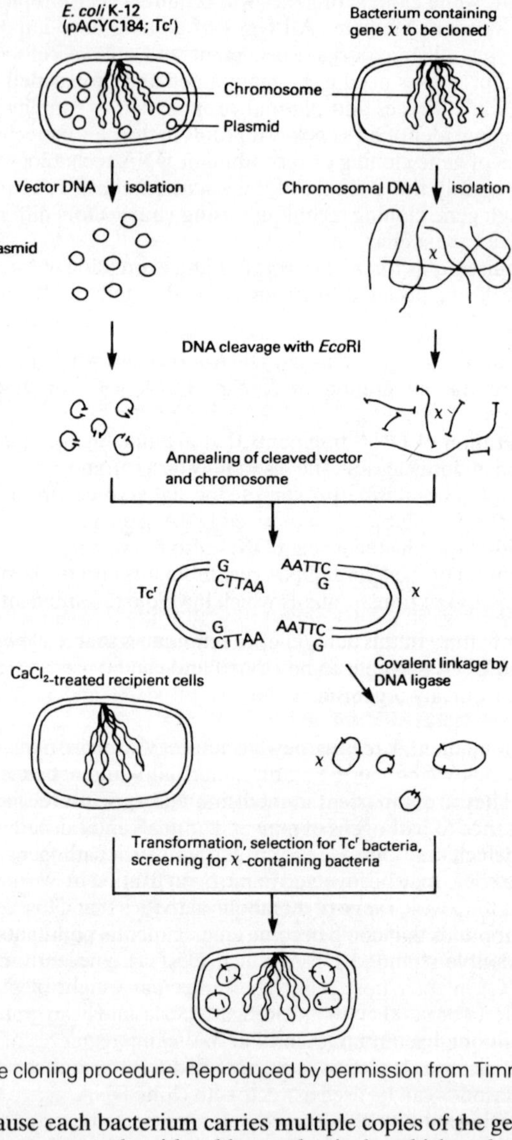

Fig. 8 The gene cloning procedure. Reproduced by permission from Timmis (1981).

products because each bacterium carries multiple copies of the genes of interest. When cells are treated with chloramphenicol, which selectively inhibits chromosomal DNA replication, there is a further substantial selective amplification of plasmid DNA.

The pBR322 plasmid contains replication determinants of the Col E1-related plasmid pMB1, the ampicillin resistance gene of plasmid R1 (Tn3) and the tetracycline resistance gene of plasmid pSC101 as selection markers. It contains single cleavage sites for at least ten endonucleases.

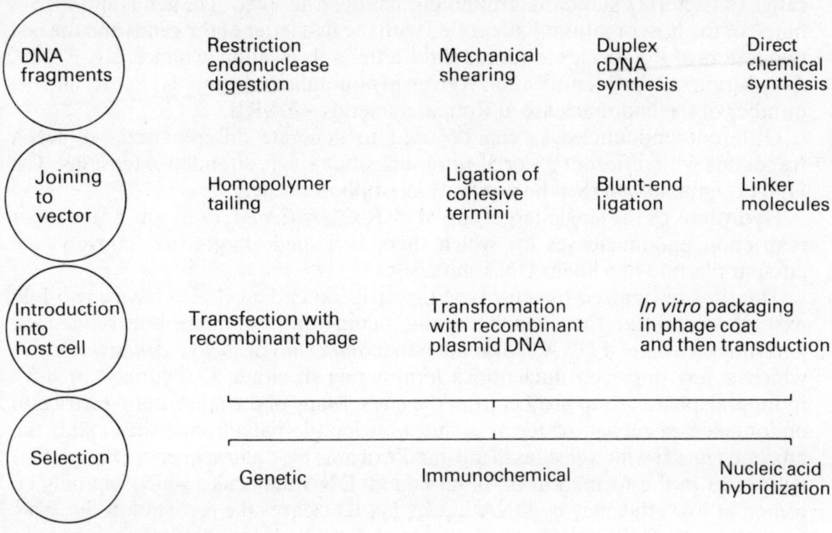

Fig. 9 Cloning strategies.

2 Restriction endonucleases Microorganisms are able to distinguish between their own DNA and that which is derived from some other source. The system that has been evolved to achieve the distinction between self and non-self relies on two types of enzyme within the cell: modification methylases and restriction endonucleases. The methylases add methyl groups in a specific fashion to various nucleotide residues in the nascent double-stranded DNA chain. Such recognition sequences are normally from four to seven base pairs in length and are palindromic, that is, the sequence in one strand of the DNA duplex is the same as in the complementary strand. Most organisms possess a number of different modification methylases. The restriction endonucleases identify the same sequences as the modifying enzyme but, instead of methylating, cut the DNA at the sequence. However, the endonuclease will not cut 'self' sequences but will degrade 'foreign' DNA. This property of recognition and cleavage of specific sequences of double-stranded DNA has become the basis for excising DNA fragments from most types of organisms.

Different types of restriction endonucleases occur, namely types I, II, III and IV. Only type II is of importance to the new DNA technology.

Type II restriction endonucleases have special importance because they are sequence specific and can create double-stranded breaks at exact locations in double-stranded DNA. In this way they break up long foreign DNA molecules into smaller segments as well as make a single break in the plasmid vector at the site at which DNA fragments can be inserted. With some of these enzymes, the breaks produce single-stranded ends and the base sequence of these termini will be identical for all fragments produced by a given restriction nuclease. In this manner any fragment generated by a specific endonuclease can base-pair with any other fragment cleaved by the same endonuclease and anneal thus creating a hybrid molecule. Since many restriction endonucleases are now available from a range of bacteria a standard terminology has been adopted. The genus and species name of the host organism is identified with the first letter of the genus and the first two letters of the species to form a three letter abbreviation in italics, e.g. *E. coli*, *Eco*. Strain or type identification is given in non-italicized symbols (*Eco*R) and the number of the endonuclease in Roman numerals – *Eco*RII.

Different endonucleases can be used to generate different sizes of DNA fragments with different 3′- or 5′-tetranucleotide single-stranded extensions. The DNA fragments can then be purified electrophoretically.

Hydrolysis of the single target site of pBR322 by *Eco*RI, or by any of the other restriction endonucleases for which there is a single target site, converts the circular plasmid to a linear DNA molecule.

The sites recognized by most type II restriction endonucleases have a two-fold axis of symmetry: they can be tetra-, penta- or hexa-nucleotide sequences. Enzymic cleavage of DNA can create either *blunt ends* or *sticky, 'cohesive ends'* in which a few unpaired nucleotides terminating in either 3′- hydroxyl or a 5′-monophosphate group project from the ends. Many of the most useful restriction endonucleases recognize tetra- or hexa-nucleotide palindromic sites. Thus the enzyme *Hinc* II which cleaves in the middle of a six base pair sequence (Figure 10a) will result in the formation of 'blunt ended' DNA molecules which can only be joined at low efficiency by DNA ligase. *Eco*RI cleaves the recognition site away from the centre (Figure 10b) and because of the palindromic nature of the enzyme recognition site causes the formation of double-stranded DNA fragments with single stranded termini or 'sticky ends'. When mixtures of these fragments are mixed hydrogen bonding occurs between complementary bases on the single stranded sequences and the ends become associated.

A particular hexanucleotide sequence should occur on average every 4096 (4^6) base pairs. Fragments generated by an endonuclease recognizing a tetranucleotide sequence would have an average length of 256 base pairs: note that the amount of DNA required to code for a typical protein of molecule mass 35 000 would be about 1000 nucleotides long.

DNA fragments for cloning may also be obtained other than by using restriction endonucleases: these include mechanical shearing, sonication, chemical synthesis, or by using the enzyme reverse transcriptase to synthesize 'copy DNA' (c DNA) from a eukaryotic messenger RNA template. This last method overcomes the problem of *introns* or *intervening sequences* which are characteristic of most genes in higher eukaryotes. Such introns cannot be removed during transcription of eukaryotic DNA in bacteria whereas in eukaryotic expression they are removed by *splicing* the primary RNA transcript of the gene.

3 Joining to vectors The insertion and linking of DNA fragments into plasmid vectors can be achieved by several methods.

Treatment of mixtures of DNA fragments with *E. coli* or bacteriophage T4 DNA ligase will generate covalent linkages between the cloning vector and the DNA molecules. Covalent phosphodiester bonds will be formed between the fragments which have annealed to each other by their cohesive ends.

A much less efficient method of ligation is 'blunt end' or 'flush-end' ligation (Figure 10a). In this case the DNA molecules do not have projections of single-stranded DNA and require high concentrations of DNA ends and enzymes for effective ligation. However, this method does allow a much wider range of endonucleases to be used increasing the range of DNA under study. When such a

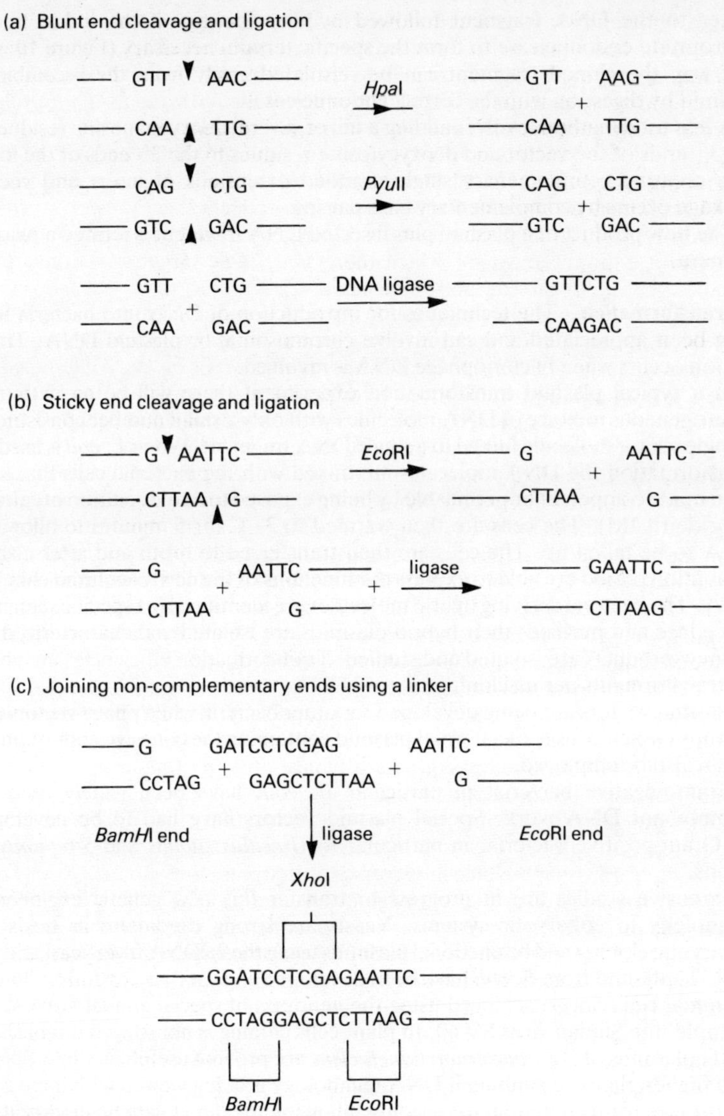

Fig. 10 Mode of action of restriction and ligase enzymes.

hybrid has been formed it cannot be cleaved by either of the endonucleases which were used to generate the original DNA fragments.

This potential disadvantage can be overcome by using *linkers*, i.e. short synthetic polynucleotides containing one or more endonuclease recognition sequences. Many linkers are now available which allow easy modification of DNA fragment termini. The use of linker sequences permits blunt-end ligation of the linker to the DNA fragment followed by the cleavage of the linker by the appropriate endonuclease to form the specific termini necessary (Figure 10c). In this way the cloned fragment can be reisolated easily from the recombinant plasmid by digestion with the correct endonuclease.

A less used method involves adding a tail of several deoxyguanosine residues to the 3′- ends of the vector and deoxycytosine residues to the 3′- ends of the insert thus creating complemenary single-stranded extensions of insert and vector. Linkage occurs by complementary base pairing.

The new product of a plasmid plus inserted DNA fragment is termed a plasmid chimera.

4 Transformation The techniques for introduction of DNA into bacteria have long been appreciated and can involve chromosomal or plasmid DNA. Transfection occurs when bacteriophage DNA is involved.

In a typical plasmid transformation experiment there will be an extremely heterogeneous mixture of DNA molecules with only a small number consisting of a single vector molecule linked to a single DNA fragment. When *E. coli* is used for transformation the DNA molecules are mixed with the bacterial cells that have been made competent or permeable by being exposed to a cold solution of calcium chloride (0.1M). The cells are then warmed to 37°C for 5 minutes to allow the DNA to be taken up. The cells are then transferred to broth and after a short incubation period are able to express the functions of the newly acquired chimeric DNA. The bacteria carrying hybrid molecules are identified by a specific screening procedure and purified: their hybrid plasmids are isolated and characterized, or the new products are isolated and studied. Transformation efficiencies are about 10^6 transformants per μg cloning mixture DNA.

Plasmid vectors are being developed for other bacteria while phage vectors and cosmids (which consist of a normal plasmid containing the cohesive ends of phage λ) can also be employed.

Gram-negative bacteria, in particular *E. coli*, have been widely used for recombinant DNA work. Special plasmid vectors have had to be developed for Gram-positive bacteria, in particular for *Bacillus subtilis* and *Streptomyces* strains.

Extensive studies are in progress to transfer this new genetic engineering technology to eukaryotic systems. Yeasts are strong candidates as hosts for eukaryotic cloning and bifunctional plasmids using the 2μDNA from yeast and the pBR322 plasmid from *E. coli* have been developed and used successfully. Cloning in animal cells can be achieved using the genomes of special animal viruses, for example, the Simian virus SV 40. In plant cells cloning is not so well established; the Ti plasmids of *Agrobacterium tumefaciens* are proving useful.

In higher plants recombinant DNA technology is being viewed with increasing importance for its potential in breeding, allowing the direct genetic modification of useful plants to improve their productivity. The functional expression of a novel gene in a genetically engineered plant has yet to be reported. A major reason for this is the limited understanding of the molecular basis of gene expression in plants. Furthermore, the genes for most important plant characteristics have yet to be identified. Whereas populations of engineered cells are the useful product in microbial genetic engineering, engineered plant cells are of little value if they cannot be regenerated into whole plants. Unfortunately, relatively few agronomically significant crop plants can yet be regenerated from cell cultures. The potential for the improvement of crop plants through genetic engineering is undoubtedly vast but more basic research is required in almost all areas of plant molecular biology before it can be realized.

5 Identification of transformants In most cloning experiments the major part of the programme will be concerned with the identification of the transformed recombinant clones. Normally the quantity of transforming DNA incorporated into the recipient host will be very much less than the host chromosomal DNA, so identification of product expressed from the transforming fragment will be extremely difficult. However, various methods have been developed to identify clones that carry desired DNA sequences.

(a) Complementation Some in-coming genes are able to complement defective copies of the same gene on the chromosome of the recipient: that is, they code for a normal, active product for which the recipient is deficient. For example, the *lacz*+ gene (coding for β-galactosidase) will complement a *lacz*, β-galactosidase-deficient host to yield lac^+ recipients which can be selected as red colonies on MacConkey's Agar.

(b) Antibiotic resistance genes These genes confer resistance to an antibiotic in normally susceptible cells. This can be exemplified with plasmid pBR322 which carries genes for tetracycline and ampicillin resistance and transformants can be selected on media containing one of these antibiotics: the foreign DNA is usually cloned into the gene coding for resistance to the other, unselected antibiotic.

(c) Hybridization A widely used method of analysing a collection of recombinant DNA molecules or cloned cells is by colony or plaque hybridization. Cells containing recombinant DNA are transferred to a nitrocellulose filter and lysed. Purified, complementary radioactive nucleic acids are then hybridized to the DNA in the nitrocellulose-bound cells. Complementary DNA will hybridize and it can be detected by autoradiography. The extent of hybridization will reflect the amount of complementarity. Copy DNA synthesized from an mRNA template can also be used to screen for recombinant DNA bound to a nitrocellulose filter in a similar way.

(d) Restriction endonuclease analysis This method can be used to determine whether the cloned DNA produces the same size of fragments as a standard DNA molecule. Restriction endonuclease analysis provides mapped fragments which are often a prerequisite to sequence analysis.

(e) Immunological methods Immunological methods are finding increasing use and depend largely on the development of antibodies, primarily against protein products.

Identification of recombinant DNA is a highly secretive aspect of industrial cloning programmes and the results of most of the studies will remain privileged information. This is an inevitable but regrettable aspect of the commercial importance of these new technologies.

Genetic engineering must be properly seen as the capstone of the massive framework of applied genetics which has been developed so successfully over the past three decades. Notwithstanding, it is certainly the most exciting and potentially the most creative technique available to the industrial geneticist.

The scientific and economic applications of recombinant DNA technology are limitless and will include:

(i) Increasing the yield of a particular gene product.
(ii) Improving the rate of synthesis of a particular end product by cloning additional enzymes into the organism or by specifically altering a cloned enzyme by site-directed mutation (e.g. cloning energy-conserving nitrogen pathway from *E. coli* into ICI single cell protein organism *Methylophilus methylotrophus*).
(iii) 'Tailoring' an organism by transferring a particular activity to a more desirable host organism.

Recombinant DNA technology will have a major impact on pharmaceutical products in the very near future. It is now possible to introduce genes from mammalian sources into bacteria and obtain the production on a commercial scale of the products encoded by these mammalian genes. Compounds of particular importance include insulin, human growth hormone, enzymes, antibodies, interferon and vaccines. The time scale for development and commercial application hopefully will permit many of these products to be available later this decade. In the veterinary field new animal vaccines derived by genetic engineering will soon be available. Although cloning of foreign genes is now routinely performed with *E. coli*, a major problem is that this organism secretes very few proteins. Secreted proteins are characterized by signal sequences which consist mainly of hydrophobic amino acid residues which determine the export of the protein across the cell membrane (Priest, 1984). During the export process the signal peptide is cleaved off to yield the protein or hormone. In part this problem with *E. coli* can be overcome by fusing the gene for the desired product with another foreign gene whose gene product is exported; another approach would be to make the cell more 'leaky'. However, other bacteria such as *Bacillus subtilis*, which is becoming increasingly used in genetic engineering and can excrete over 50 proteins, will undoubtedly find greater use in industrial application of foreign gene cloning.

In the long term the greatest potential for recombinant DNA techniques should be in agriculture. The possibility of transferring the nitrogen fixation genes from bacteria into important crop plants other than legumes is being actively studied. Similarly the possibility of transferring the structural genes of legumes controlling nitrogen-fixing association with *Rhizobium* is also being studied.

In the future many other areas of biotechnology including energy and chemical feedstocks and waste treatment will benefit from the ability of applied geneticists to produce the 'ideal' organism for a desired purpose.

Although there are now many genetic manipulations that can be carried out with organisms it is worth noticing that the following important technical limitations presently restrict wide industrial applications.

(1) There is inadequate information about (or concerning) the chromosome location of desired genes for most industrially useful organisms, e.g. *Penicillium chrysogenum, Aspergillus oryzae*.
(2) The availability of useful vectors for most industrially important organisms is at an early stage of development.
(3) Physiological patterns and, in particular, enzymic steps catalysing

important conversions are not well understood for many important processes.

(4) Processes that require multiple gene involvement present formidable problcms.

(5) Chimeric plasmids will only be useful in host cells if they are stably retained within the cell. The danger is not that the plasmid will be broken up but that after a few generations in continuous culture it may simply be lost. The additional 'genetic load' of a chimeric plasmid may affect the growth rate and chimeric plasmid-free cells may outgrow and replace the other cells. However, many studies are now showing that instability amongst plasmids can be overcome either by novel changes or insertions in the chimeric plasmid or by careful control of fermentation parameters.

Regulation and control of recombinant DNA experimentation

The ability of recombinant DNA technology to produce hybrid molecules of DNA never before seen in nature has caused concern within society (Table 8). In an industrial context genetic manipulations by the new techniques have proceeded rapidly and are now widespread. Advances in this technology can be translated into economic power and its marketability may obscure the profound ecological and ethical implications of genetic engineering.

The original guidelines set out by countries involved in genetic engineering were initially very restrictive but have now been relaxed to the point where they afford little protection. Although the early fears about the risks involved in genetic manipulation were exaggerated, constant vigilance should be maintained.

Producing new strains of organisms for biotechnological processes can be extremely costly and for this reason most industrial concerns keep a high security level around important producer strains. Although living organisms can now be patented in many countries there is considerable doubt that patents can offer worthwhile protection. *The Budapest Treaty on the International Recognition of the Deposit of Microorganisms for the Purpose of Patent Procedure (1978)* legitimizes existing national practices of countries that have signed the Convention and allows for international recognition, through the issue of a patent, of a microorganism which has been transmitted to the depository. How successful this treaty will be has yet to be shown.

Summary

Applied genetics is concerned with deriving new and improved strains of organisms that can be utilized for the benefit of mankind. In biotechnology this will involve the selection of organisms from natural sources, from culture collections and other organizations, or by further exploitation of 'in house' strains. A wide variety of techniques is available to modify, delete or add to the genetic complement of organisms. Selection and screening activities occupy a major part of biotechnological programmes. Screening is the use of procedures to allow the detection and isolation of only those organisms or metabolites of interest among a large population. Screens may be non-selective random screens in which all the

Table 8 Major events in the commercialization of new biotechnology

1973	First gene cloned
1974	First expression of a gene cloned from a different species in bacteria.
	Recombinant DNA (rDNA) experiments first discussed in a public forum (Gordon conference).
1975	US guidelines for rDNA research outlined (Asilomar conference).
1976	First firm to exploit rDNA technology founded in the United States (Genetech).
	Genetic Manipulation Advisory Group (UK) started in the United Kingdom.
1980	Diamond v. Chakrabarty – US Supreme Court rules that microorganisms can be patented under existing law. Cohen/Boyer patent issued on the technique for the construction of rDNA.
	United Kingdom targets biotechnology (Spinks' report).
	Federal Republic of Germany targets biotechnology (Leistungsplan).
	Initial public offering by Genetech sets Wall street record for fastest price per share increase ($35 to $89 in 20 minutes).
1981	First monoclonal antibody diagnostic kits approved for use in the United States.
	First automated gene synthesizer marketed.
	Japan targets biotechnology (Ministry of International Trade and Technology declares 1981 'The Year of Biotechnology').
	France targets biotechnology (Pellsolo report).
	Hoesch/Massachusetts General Hospital agreement.
	Initial public offering by Cetus sets Wall street record for the largest amount of money raised in an initial public offering ($125 million).
	Industrial Biotechnology Association founded.
	DuPont commits $120 million for life sciences R & D.
	Over 80 new biotechnology firms (NBFs) had been formed by the end of the year.
1982	First rDNA animal vaccine (for collbacillosis) approved for use in Europe.
	First rDNA pharmaceutical product (human insulin) approved for use in the United States and the United Kingdom.
	First R & D limited partnership formed for the funding of clinical trials.
1983	First plant gene expressed in a plant of a different species.
	$500 million raised in US public markets by NBFs.

Source: Office of Technology Assessment.

isolates are treated as individuals for the requisite qualities or they may be rational screens involving some aspect of preselection.

Producer organisms must be preserved with minimum degeneration of genetic capabilities. Preservation can be on agar medium, by reduced metabolism, drying, freeze drying or by ultra-low temperature.

Organism genomes can be modified by mutagenesis or by various forms of hybridization. Mutational programmes are primarily aimed at strain improvement and mutagens available include UV and ionizing radiation, thymine starvation and a wide selection of chemical mutagens. The mutation rate in an organism is genetically controlled and can be altered by mutator or anti-mutator genes.

Hybridization is essentially a method that facilitates the recombination of genetic material between organisms and can be expressed by sexual or parasexual mechanisms. Sexual hybridization occurs between haploid nuclei and will involve karyogamy, nuclear fusion and finally meiosis. Recombination of the genetic characteristics will arise because of rearrangement and reorganization of the chromosomes.

Parasexual processes use various cellular mechanisms to bring together genetic material from different genetic sources. The main methods practised industrially involve conjugation, transduction, transformation and mitotic recombination.

Protoplast fusion techniques are widely practised with many microbial cells as well as with plant and animal cells. Fusion rates have been greatly increased by means of the fusogen polyethylglycol.

Recombinant DNA technologies allow the isolation, purification and selective amplification in specific host cells of discrete DNA fragments or genes from almost any organism. The basic requirements of this technology are restriction endonuclease enzymes which can cleave double-stranded DNA at specific sites generating DNA fragments of defined sizes. These fragments can then be inserted into carrier vehicles such as plasmids or phage and be covalently bonded by DNA ligase enzymes. Transformation is the main method used for the insertion of these chimeric DNA molecules into host cells. The types of host cell are typically but not exclusively specific mutant cells of *E. coli, B. subtilis* or *S. cerevisiae*. Transformed cells are identified by several means including complementation and resistance markers, hybridization, immunological methods and restriction endonuclease analysis.

References

ANON (1984). *Genetic Engineering of Plants*. National Academy Press, Washington.

CRICK, F. (1979). Split genes and RNA splicing. *Science* 204: 264–71.

DIETZ, A. (1981). Pure culture methods for industrial microorganisms. In: *Biotechnology*, Vol. I, *Microbial Fundamentals*, pp. 411–34. Edited by H. J. Rehm and G. Reed. Verlag Chemie, Basel.

ESSER, K. and STAHL, U. (1981). Hybridization. In: *Biotechnology*, Vol. I, *Microbial Fundamentals*, pp. 305–29. Edited by H. J. Rehm and G. Reed. Verlag Chemie, Basel.

EVANS, D. A. (1983). Agricultural applications of plant protoplast fusion. *Biotechnology* May, 253–61.

GLOVER, D. M. (1980). *Genetic Engineering – Cloning DNA*. Chapman and Hall, London.

HARDY, K. (1981). *Bacterial Plasmids*. Aspects of Microbiology 4. Van Nostrand Reinhold (UK), Wokingham.

HOPWOOD, D. A. (1981). The genetic programming of industrial microorganisms. *Scientific American*, 245: 90–112.

Impacts of Applied Genetics: Microorganisms, Plants and Animals (1981). Congress of the United States: Office of Technology Assessment, Washington DC.

Industrial Microbiology and the Advent of Genetic Engineering (1981). A Scientific American Book. Freeman, New York.

KIRSOP, B. (1983). Culture collections – their services to biotechnology. *Trends in Biotechnology*, 1: 4–8.

KOHLER, G. and MILSTEIN, C. (1975). Continuous cultures of fused cells secreting antibody of predefined specificity. *Nature* (London) 256: 495–7.

KORNBERG, A. (1980). *DNA Replication*, W. H. Freeman, San Francisco.

MANNIATIS, T., FRITSCH, E. F. and SAMBROOK, J. (1982). *Molecular Cloning: A laboratory manual*. Cold Spring Harbor Laboratory, New York.

MURRAY, K. (1980). Genetic engineering: possibilities and prospects for its application in industrial microbiology. *Philosophical Transactions of the Royal Society, London* 290: 369–89.

NISBET, L. J. (1982). Current strategies in the search for bioactive microbial metabolites. *Journal of Chemical Technology and Biotechnology* 32: 251–70.

OLD, R. W. and PRIMROSE, S. B. (1981). *Principles of Gene Manipulation: An Introduction to Genetic Engineering*. Blackwell, London.

PEBERDY, J. F. (1979). Fungal protoplasts: isolation, reversion and fusion. *Annual Review of Microbiology* 33: 21–39.

PRIEST, F. G. (1984). *Extracellular Enzymes*. Aspects of Microbiology 9. Van Nostrand Reinhold (UK), Wokingham.

ROWLANDS, R. T. (1983). Industrial fungal genetics and strain selection. In: *The Filamentous Fungi*, Vol. 4, *Fungal Technology*, pp. 346–72. Edited by J. E. Smith, D. R. Berry and B. Kristiansen. Edward Arnold, London.

SCHELL, J. and VAN MONTAGU, M. (1983). The Ti plasmids as natural and as practical gene vectors for plants. *Biotechnology* April, 175–80.

SECHER, D. S. (1980). Monoclonal antibodies by cell fusion. *Immunology To-day* July, 22–6.

SETLOW, J. K. and HOLLAENDER, A. (Eds) (1980). *Genetic Engineering: Principles and Methods*. Plenum Press, New York and London.

TIMMIS, K. N. (1981). Gene manipulation *in vitro*. *Symposium Society of General Microbiology* 31: 49–109.

WATSON, J. D., TOOZE, J. and KURTZ, D. T. (1984). *Recombinant DNA: a short course*. A Scientific American Book. Freeman, New York.

WILSON, T. (1984). Engineering to-morrow's vaccines. *Biotechnology* January, 29–39.

ZIMMERMANN, U. (1983). Electrofusion of cells: principles and industrial potential. *Trends in Biotechnology* 1, 149–55.

3 Fermentation technology

The nature of fermentation

The origins of fermentation technology were largely with the use of microorganisms for the production of foods and beverages such as cheeses, yoghurts, alcoholic beverages, vinegar, sauerkraut, fermented pickles and sausages, soya sauce and the products of many other Oriental fermentations (Table 9). The present-day large scale production processes of these products are essentially scaled up versions of former domestic arts. Paralleling this development of product formation was the recognition of the role microorganisms could play in removing unpleasant wastes and this has resulted in massive world-wide service industries involved in water purification, effluent treatment and waste management. New dimensions in fermentation technology have made use of the ability of microorganisms (1) to overproduce specific essential primary metabolites such as glycerol, acetic acid, lactic acid, acetone, butyl alcohol, butane diol, organic acids, amino acids, vitamins, polysaccharides and xanthans; (2) to produce useful secondary metabolites (groups of metabolites that do not seem to play an immediate recognizable role in the life of the microorganism producing them) such as penicillin, streptomycin, oxytetracycline, cephalosporin, giberellins, alkaloids, actinomycin; and (3) to produce enzymes as the desired industrial product such as the exocellular enzymes amylases, proteases, pectinases or intracellular enzymes such as invertase, asparaginase, uric oxidase, restriction endonucleases and DNA ligase. More recently, fermentation technology has begun to use cells derived from higher plants and animals under conditions known as cell or tissue culture. Plant cell culture is mainly directed towards secondary product formation such as alkaloids, perfumes and flavours while animal cell culture has initially been mainly concerned with the formation of protein molecules such as interferons, monoclonal antibodies and many others.

Future markets are largely assured for fementation products because, with limited exceptions, it is not possible to produce them economically by other chemical means. Furthermore economies in production will also occur by genetically engineering organisms to unique or higher productivities. The commercial market for products of fermentation technology is almost unlimited but will ultimately depend on economics and safety considerations.

The processes of commercial fermentation are in essence very similar no matter what organism is selected, what medium is used and what product formed. In all cases, large numbers of cells with uniform characteristics are grown under defined, controlled conditions. The same apparatus with minor modifications can be used to produce an enzyme, an antibiotic, an organic chemical or a single cell protein. In its simplest form fermentation processes can be just the mixing of microorganisms with a nutrient broth and allowing the components to react. More advanced and sophisticated large-scale processes require control of the entire environment so that the fermentaton process can proceed efficiently and, what is more important, can be exactly repeated with the same amounts of raw

Table 9 Fermentation products according to industrial sectors

Sector	Activities
Chemicals	
Organic (bulk)	Ethanol, acetone, butanol
	Organic acids (citric, itaconic)
Organic (fine)	Enzymes
	Perfumeries
	Polymers (mainly polysaccharides)
Inorganic	Metal beneficiation, bioaccumulation and leaching(Cu, U)
Pharmaceuticals	Antibiotics
	Diagnostic agents (enzymes, monoclonal antibodies)
	Enzyme inhibitors
	Steroids
	Vaccines
Energy	Ethanol (gasohol)
	Methane (biogas)
	Biomass
Food	Dairy products (cheeses, yogurts, fish and meat products)
	Beverages (alcoholic, tea and coffee)
	Baker's yeast
	Food additives (antioxidants, colours, flavours, stabilizers)
	Novel foods (soy sauce, tempeh, miso)
	Mushroom products
	Amino acids, vitamins
	Starch products
	Glucose and high fructose syrups
	Functional modifications of proteins, pectins
Agriculture	Animal feedstuffs (SCP)
	Veterinary vaccines
	Ensilage and composting processes
	Microbial pesticides
	Rhizobium and other N-fixing bacterial inoculants
	Mycorrhizal inoculants
	Plant cell and tissue culture (vegetative propagation, embryo production, genetic improvement)

Adapted from Bull *et al.* (1982).

materials, broth and cell inoculum producing precisely the same amount of product.

All biotechnological processes are carried out within a containment system or bioreactor. The physical form of most common bioreactors has not altered much over the past 30 years. Recently, however, many novel forms have been designed and they may play an increasingly active part in biotechnology. The main function of a bioreactor is to minimize the cost of producing a product or service while the driving force behind the continued improvement in design and function has been the need to increase the rate of product formation and the quality of the product or service. Studies have considered better aseptic design and operation, better process control including computer involvement, and how to obtain a better understanding of the rate-limiting steps of a system, especially heat and mass transfer.

In biotechnology, processes can be broadly considered to be either conversion cost intensive or recovery cost intensive. With conversion cost intensive processes the volumetric productivity, Qp (kg of product $m^{-3}h^{-1}$), is of major importance while with recovery cost intensive processes the product concentration, P (kg m^{-3}), is the main criterion for the minimization of cost. Examples of the diverse product categories produced in bioreactors in the biochemical process industry are given in Table 10 while Table 11 identifies the various cultivation methods employed in biotechnology.

Table 10 Examples of products in different categories in biotechnological industries

Category	Example
Cell mass*	Baker's yeast, single cell protein
Cell components†	Intracellular proteins
Biosynthetic products†	Antibiotics, vitamins, amino and organic acids
Catabolic products*	Ethanol, methane, lactic acid
Bioconversion*	High-fructose corn syrup, 6-aminopenicillanic acid
Waste treatment	Activated sludge, anaerobic digestion

* Typically conversion of feedstock cost-intensive processes.
† Typically recovery cost-intesive process.
From Cooney (1983).

There are three main operating types of bioreactors for biotechnological processes together with two forms of biocatalysts. Bioreactors can be operated on a batch, semi-continuous (fed-batch) or continuous basis. Reactions can occur in static or agitated cultures, in the presence or absence of oxygen, and in aqueous or low moisture conditions (solid substrate fermentations). The biological catalysts can be free or can be attached to surfaces by immobilization or by natural adherence. The biocatalysts can be cells in a growing or non-growing state or isolated enzymes used as soluble or immobilized catalysts. In general, the reactions occurring in a bioreactor are conducted under moderate conditions of pH (near neutrality) and temperature (20 to 65°C). In most bioreactors the processes occur in an aqueous phase and product streams will be relatively dilute.

The optimization of a bioreactor process involves minimizing the use of raw materials (e.g. nutrients, precursors, acid/base, air) and energy (the cost of energy since 1978 has risen at an annual rate of 16%), and maximizing product purity and quality in the broth before recovery. Process optimization is achieved by manipulation of both the physical and chemical parameters associated with the process. The range of process variables that are important to process development are listed in Table 12 and are discussed later.

The remainder of this chapter will be concerned with the principles of organism growth in bioreactors and special attention is given to microbial, plant and animal cells used in product formation. The use of bioreactors for enzymic function either in soluble or immobilized form together with certain types of immobilized microbial cell systems will be considered in Chapter 4.

Table 11 Characteristics of cultivation methods

Type of culture	Operational characteristics	Application
Solid	Simple, cheap, selection of colonies from single cell possible; process control limited	Maintenance of strains, genetic studies; production of enzymes; composting
Film	Various types of bioreactors: trickling filter, rotating disc, packed bed, sponge reactor, rotating tube (see chapter 4)	Waste water treatment, monolayer culture (animal cells); bacterial leaching; vinegar production
Submerged homogeneous distribution of cells: batch	'Spontaneous' reaction, various types of reactor: continuous stirred tank reactor, air lift, loop, deep shaft, etc.; agitation by stirrers, air, liquid process control for physical parameters possible; less for chemical and biological parameters	Standard type of cultivation: antibiotics, solvents, acids, etc.
Fed-batch	Simple method for control of regulatory effects, e.g. glucose repression	Production of baker's yeast
Continuous one-stage homogeneous	Proper control of reaction; excellent tool for kinetic and regulatory studies; higher costs for experiment; problem of aseptic operation, the need for highly trained operators	Few cases of application in industrial scale: production of single cell protein; waste water treatment

From Fiechter (1981).

Principles of microbial cultivation in aqueous systems

Growth of organisms may be considered as the increase of cell material expressed in terms of mass or cell number and is the result of a highly coordinated series of enzymically catalysed biological steps. Optimal expression of growth will be dependent on the transport of necessary nutrients to cell surfaces (mass transfer) and on environmental parameters such as temperature and pH being optimally maintained.

The quantity of cell material (X) or biomass in a bioreactor can be determined gravimetrically (by dry weight, wet weight, DNA, or protein) or numerically (by number of cells). Doubling time (t_d) relates to the period of time required for the doubling in weight of biomass while generation time (g) relates to the period necessary for the doubling of cell numbers. During balanced or exponential growth, when growth is controlled only by intrinsic cellular activities, $g = t_d$ provided every cell of the population is able to divide. Average doubling times increase with increasing cell size and complexity; the following range of values in hours has been achieved experimentally: bacteria 0.25 to 1, yeast 1.15 to 2, mould fungi 2 to 6.9 and plant cells 20 to 40.

Table 12 Process variables in fermentation processes.

Temperature	pH	Respiratory quotient
Pressure	Oxidation reduction potential	Cell concentration
Agitation speed	Dissolved oxygen	Cellular composition:
Power input	Dissolved CO_2	protein (enzymes)
Air flow rate	Effluent oxygen	DNA
Feed rate of:	Effluent CO_2	RNA
nutrients	Dissolved concentations of:	lipid and carbohydrate
precursors	carbohydrate	Specific activity of
inducers	nitrogen	enzyme
Weight of broth	mineral ions	Specific rates of:
Volume of liquid	precursor	product formation
Viscosity (apparent)	inducer	growth
Cumulative amount of:	product	oxygen uptake
acid	metabolites	precursor utilization
base	Broth rheology	nutrient uptake
antifoam	Power characteristics	CO_2 production
	Energy balance	Oxygen transfer rate

Adapted from Ryu and Humphrey (1973) *Journal of Applied Chemistry and Biotechnology* 23, 283–95.

Under ideal conditions the biosynthetic potential of microorganisms can be very great with a biomass doubling time of only 15 minutes for certain types of bacteria. However, it is seldom that optimal growth conditions will prevail for any length of time and in practice growth will become dependent on a limiting factor, for example, an essential nutrient. As the concentration of this factor drops to zero, so also will the growth potential of the organism decrease. The classical studies of Monod (1942) have allowed the development of mathematical equations describing the essential features of microbial growth in bioreactors. The original mathematical equation described the specific growth rate μ as a function of the concentration S:

$$\mu = \mu_{max} \frac{S}{K_S + S} \tag{1}$$

In this case S is the concentration of a substrate in the medium which is in limiting concentration when compared with other essential nutrients, μ_{max} is the maximum specific growth rate of the organism while K_S represents a saturation constant. K_S is the substrate concentration at which $\mu = \mu_{max}/2$. Thus exponential growth can occur at specific growth rates having any value between zero and μ_{max} if the substrate concentration can be held constant at the appropriate value (which will be of significance for continous culture). Identification of the nutrients essential for growth as well as the optimum conditions required for growth have been derived from batch and continuous bioreactor systems. The rate of increase of concentration of organisms (dx/dt) is the *growth rate* while the *specific growth rate* is the rate of increase/unit of organism concentration $(1/x)$ (dx/dt). A simple relationship exists between growth and utilization of substrate. In simple systems growth rate is a constant fraction, Y, of the substrate utilization rate:

$$\frac{dx}{dt} = -Y\frac{ds}{dt} \tag{2}$$

Y is the *yield constant* and over any finite period of growth

$$Y = \frac{\text{weight of cells formed}}{\text{weight of substrate used}}$$

Knowing the values of the three growth constants μ_{max}, K_S and Y, equations (1) and (2) can give a complete quantitative description of the growth cycle of a batch culture.

In batch culture the inoculum and nutrients that will be required for growth will be brought together in a containment vessel under optimum environmental conditions of temperature, pH and mixing. This represents a closed system except for aerobic organisms where a continuous supply of air will normally be supplied to the bioreactor.

In batch cultivation, the growth rate and specific growth rate will seldom be constant, reflecting the changing nutrient environment characteristic of the system. The complex nature of batch growth of microorganisms is shown in Figure 11. The initial *lag phase* is a period of no observable growth but chemical analyses show much hidden metabolic turnover indicating that the cells are adapting to the

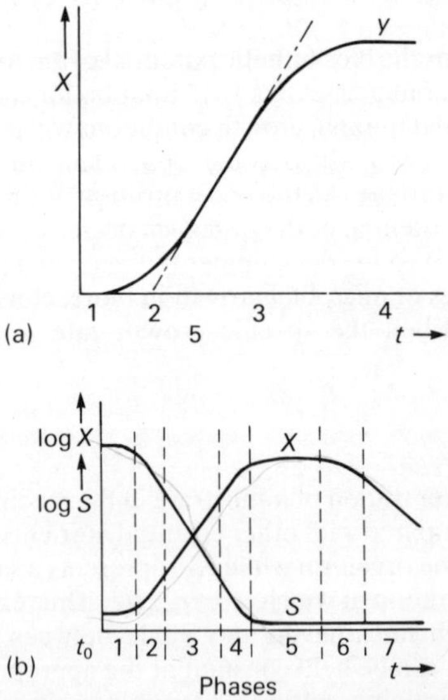

Fig. 11 Growth characteristics in a batch culture of a microorganism. (a) Biomass concentration (or mass) X as a function of time: 1. Lag phase: synthesis of mRNA and enzymes, any production of biomass cancelled by destruction, zero net growth (μ=0). 2. Exponential growth phase, $\mu=\mu_{max}$. 3. Centre part of decelerating phase. 4. Zero net growth, μ_0=0. 5. Span of variable μ-values, $\mu_{max} \rightarrow <\mu_{max} \rightarrow 0$. (b) Logarithm of concentration of mass of X (biomass) and S (substrate, e.g. glucose) vs. time, semilogarithmic plot (t scale linear): 1. Lag phase, μ=0. 2. Accelerating phase. 3. Exponential phase, $\mu=\mu_{max}$. 4. Decelerating phase. 5. Stationary phase, μ=0 (cf. 4 in Fig. 1a). 6. Decline phase. 7. Death. Reproduced by permission from Fiechter (1981).

new environment and that growth will begin in due course. The physiological condition of the inoculum is now considered not only to be a major influencing factor of the duration of the lag phase but also to influence the nature of future growth and product formation, e.g. antibiotic synthesis. There is a *transient acceleration phase* after inoculum outgrowth and before exponential growth occurs. This phase is not well understood physiologically or mathematically because the population of cells is heterogeneous in age structure and metabolism. In the *exponential phase* organism growth is unlimited with nutrients in excess and inhibitors absent. The specific growth rate will achieve its maximum, i.e. $\mu = \mu_{max}$. However, in most batch cultivations exponential growth is transitory. As nutrients are used up by the growing cell population unlimited growth will be replaced by limited growth and although the cell population is still increasing the specific growth rate at any particular point tends to become smaller and smaller, i.e. $<\mu_{max}$. This *deceleration phase* is then followed by a *stationary phase* in which overall growth can no longer be achieved due to nutrient exhaustion. Biomass balance arises because

$$\text{rate of growth} = \text{rate of death.}$$

Many important biotechnological products are optimally formed during this period of the growth cycle, e.g. antibiotics. The final phase of the cycle is the *death phase* when the specific growth rate is negative ($\mu < 0$). Most biotechnological batch processes are stopped before this phase is reached, because of decreasing metabolism and cell lysis.

In practice, batch cultivation is used to optimize organism or biomass production and then to carry out specific chemical transformations such as end-product formation (antibiotics, organic acids) or decomposition of substances (sewage treatment). Many important products are optimally formed during the stationary phase of the growth cycle in batch cultivation.

In contrast to batch culture, in continuous cultivation the addition of nutrients and the removal of an equal fraction of the total culture volume occur continuously. Continuous methods of cultivation will permit organisms to grow under steady state conditions; that is, growth occurs at a constant rate and in a constant environment. Factors such as pH and the concentrations of nutrients and metabolic products which inevitably change during the growth cycle of a batch culture can be held constant in a continuous culture. Indeed, these parameters can be independently controlled allowing the experimenter to obtain realistic information on the role of each to the growth of the organism.

In a completely mixed continuous culture system sterile medium is fed into the bioreactor at a steady flow-rate (f) and culture broth emerges from it at the same rate keeping the volume of culture in the vessel (v) constant. With efficient mixing incoming medium should be rapidly and uniformly dispersed throughout the bioreactor. The characteristics of all stirred continuous bioreactor systems can be described mathematically by setting up balanced equations for cells, substrates, products, etc. in which the overall rate of change in concentration of any component is given as the resultant of the rates of all the processes tending to increase or decrease it. In practice these can be (a) increase due to inflow to the bioreactor—equal to the inflow rate multiplied by the concentration of component in the inflow, (b) decrease due to outflow, (c) increase (of cells) due to growth, (d) decrease (of substrates) due to utilization, (e) increase in product formation and biomass.

Residence-time, in a continuous bioreactor, is determined not by the absolute values of the flow-rate and culture volume but by their ratio, the dilution rate D, where $D = f/v$, or the number of complete volume changes per hour. The mean residence time of a particle in the culture vessel is equal to $1/D$. Assuming complete and uniform mixing every cell in the bioreactor has an equal probability of leaving it or being washed out within a given time.

The net increase of organisms is given by the simple equation

increase = growth − output

$$\frac{dx}{dt} = \mu x - Dx$$

When $\mu > D$, dx/dt is positive and cell concentration will increase; when $\mu < D$, dx/dt is negative and cell wash-out will occur; when $\mu = D$, $dx/dt = 0$ and x is constant. In this case, the steady state has been achieved where the concentration of organisms will not change with time.

Dilution rate will also affect the concentration of substrates in the bioreactor. In the bioreactor substrate enters at a concentration s_R, is consumed by the organism and flows out at a concentration s. The net rate of change of substrate concentration is obtained by another balanced equation:

increase = input − output − consumption

$$= \text{input} - \text{output} - \frac{\text{growth}}{\text{yield constant}}$$

$$\frac{ds}{dt} = Ds_R - Ds - \frac{\mu x}{Y}$$

When the dilution rate exceeds the μ_{max} value, wash-out of the organisms will occur.

When a continuous culture system is viewed as a production system (e.g. SCP process) its performance will be judged by two criteria: (1) the quantity of cells produced in unit time—*the output rate*; and (2) the quantity of cells produced from unit weight of substrate—*the effective yield* or 'yield coefficient'. In the steady state condition the total output will be equal to the product of flow-rate and concentration of organisms. For maximum output of cells or biomass the dilution rate should be high but obviously it cannot exceed μ_{max}. In practice maximum production efficiency combining a high output with efficient utilization of substrate will be obtained with a flow-rate at or a little below the maximum output rate and the highest practicable substrate concentration. Such optimum conditions will only be relevant to biomass production. Although similar conditions would be used when the desired product, for example ethanol, is a fermentation product which is formed in proportion to the amount of substrate consumed, widely different conditions might be required for the production of complex metabolic products such as antibiotics.

Fed-batch culture (see below) is a form of cultivation which involves a continuous or sequential addition of medium or substrate to the initial batch without any withdrawal from the system. Product yield from such systems can well exceed conventional batch culture. This approach is widely practised in industry—for example, in the production of baker's yeast.

In practice, batch, fed-batch and continuous cultivation systems are used in industry for production of biomass or cellular products. For many reasons batch cultivation techniques represent the dominant form of industrial practice. For a fuller understanding of the kinetics of the various techniques of growing organisms, reference should be made to Pirt (1975) and Fiechter (1981).

Bioreactor design

The bioreactor is the containment system for the biological reactions of a biotechnological process. It will provide the correct environment for optimization of organism growth and metabolic activity; it should prevent contamination of the production culture from the environment while also preventing release of the culture into the environment, and it should have the associated instrumentation or probes to allow optimum process control (Table 13).

Many bioreactor systems will be required to operate under aseptic conditions. In most systems of industrial importance pure cultures of the producer organism will be used and the presence of foreign unwanted contaminating organisms can influence the process in several ways—for example, by interfering with the biological catalyst, destroying the product, producing substances that can impair downstream processing, and introducing toxic molecules into the system.

To avoid these problems the medium, bioreactor and all ancillary pipework must be sterilized (usually with high pressure steam) and all incoming air must be passed through sterilized glass wool to remove contaminants. In batch fermentation medium is normally sterilized in the bioreactor whereas in continuous systems external sterilization is practised. In the fermentation industry there will be occasions when contaminating microorganisms do enter the bioreactor and cause damage. For this reason in the antibiotics industry bioreactors are seldom larger than 200 m^3 because of the considerable loss that will occur when contamination occurs. When continuous processes are used there is even greater need for stringent sterile operation. The anticipated greater use of genetically engineered microorganisms in industry will necessitate even more expensive exclusion techniques.

For aerobic processes the design must include a mechanism for incorporating air and mixing the contents and all systems must have provision for inoculation and

Table 13 Basic bioreactor design criteria

1. Microbiological and biochemical characteristics of the cell systems (microbial, mammalian, plant)
2. Hydrodynamic characteristics of the bioreactor
3. Mass and heat characteristics of the bioreactor
4. Kinetics of cell growth and product formation
5. Genetic stability characteristics of the cell system
6. Aseptic equipment design
7. Control of bioreactor environment (both macro- and micro- environments)
8. Implications of bioreactor design on downstream product separation(s)
9. Capital and operating costs of the bioreactor
10. Potential for bioreactor scale-up

From Glacken *et al*. (1984).

sampling and for charging and discharging the vessel. Energy input from the stirring, aeration and metabolic oxidation processes will need to be removed by cooling mechanisms. Measurement of energy input will be necessary to determine the overall mixing and aeration rates.

The materials for construction should be non-toxic, able to withstand pressurized steam and to resist chemical and electrolytic corrosion. Industrial bioreactors are usually constructed from highly polished stainless steel. Bioreactors exist in many shapes and sizes and the ratio of height to diameter (the aspect ratio) is an important working parameter.

The size of industrial bioreactors will be affected by the required product concentration and whether batch or continuous operation is chosen. Although continuous culture techniques are widely used in research they have found only limited use in industry, for example, SCP and ethanol production processes and waste water treatment. Almost all other industrial processes use batch or fed-batch cultivation methods.

The dominance of batch and fed-batch culture techniques in industry is due to some or all of the following reasons:

(a) Products may be required in relatively small quantities at any given time.
(b) Market requirements can be intermittent.
(c) Shelf life of certain products is short.
(d) High product concentration is required in broth to optimize downstream processing.
(e) Certain products are produced only during stationary phase of the growth cycle.
(f) Instability of some production strains requires their regular renewal.
(g) Continuous processes still offer many technical difficulties.

Although there are numerous designs for industrial bioreactors, the long established continuous stirred tank reactor (CSTR) or vessel continues to be the most widely used (Figure 12a). In bioreactors without mechanical agitation, e.g. tower and loop bioreactors, agitation is achieved with very high gas throughputs (Figure 12b). In large scale bioreactors of these types with aqueous fermentation broths it is now accepted that such designs can be economically competitive with mechanically agitated bioreactors. However, in all systems inceasing viscosity can create major problems with aeration as smaller bubbles tend to coalesce creating large bubbles with reduced interfacial areas.

In general, fermentation industries require a bioreactor which can meet a number of different running conditions including varying viscosity, aeration rate, intensity of agitation and fermentation volume and in practice the CSTR has become widely accepted. A further consideration in determining choice is that many industries will need to process different products in the same plant; thus a flexible system that can be easily modified will be favoured.

The basic design of the CSTR was developed during the 1940s and 1950s with the industrial production of penicillin. It is generally a fully baffled upright cylinder and the baffle width is 10% of the tank diameter (Figure 12a). Sterile air is introduced at the base of the vessel by way of an open pipe or ring sparger. The vertical shaft with overhead drive will carry one or more impellers depending on the aspect ratio. The impellers are usually spaced at intervals equivalent to one tank diameter along the shaft to avoid a swirling type of liquid movement. Flat-bladed (Rushton) turbines are used in the majority of bioreactors and usually

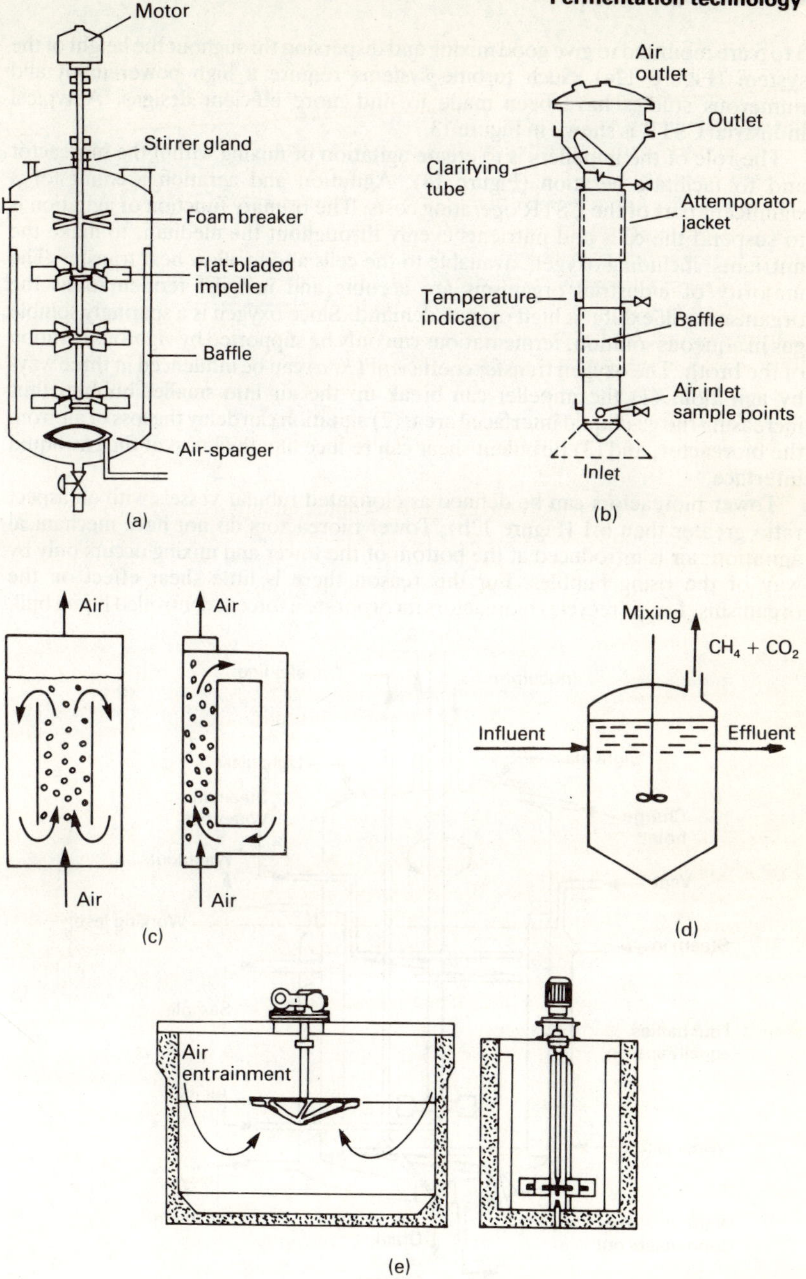

Fig. 12 Various forms of bioreactor. (a) Continuous stirred tank reactor, (b) Tower reactor, (c) Loop (recycle) bioreactor, (d) Anaerobic digester or bioreactor, (e) Activated sludge bioreactor. (a) and (b) reproduced by permission from Kristiansen and Chamberlain (1983).

3 to 5 are mounted to give good mixing and dispersion throughout the height of the system (Figure 12a). Such turbine systems require a high power input and numerous studies have been made to find more efficient designs. A typical industrial CSTR is shown in Figure 13.

The role of the impellers is to create agitation or mixing within the bioreactor and to facilitate aeration (Figure 14). Agitation and aeration account for a significant part of the CSTR operating costs. The primary function of agitation is to suspend the cells and nutrients evenly throughout the medium, to make the nutrients, including oxygen, available to the cells and to allow heat transfer. The majority of industrial organisms are aerobic and in most fermentations the organisms will exhibit a high oxygen demand. Since oxygen is a sparingly soluble gas in aqueous solution, fermentations can only be supported by vigorous aeration of the broth. The oxygen transfer coefficient ($K_L a$) can be influenced in three ways by agitation: (1) the impeller can break up the air into smaller bubbles thus increasing the gas/liquid interfacial area, (2) agitation can delay the loss of air from the bioreactor, and (3) turbulent shear can reduce film thickness at the gas/liquid interface.

Tower bioreactors can be defined as elongated tubular vessels with an aspect ratio greater than 6:1 (Figure 12b). Tower bioreactors do not have mechanical agitation; air is introduced at the bottom of the tower and mixing occurs only by way of the rising bubbles. For this reason there is little shear effect on the organisms. Loop (recycle) bioreactors incorporate a forced, controlled liquid bulk

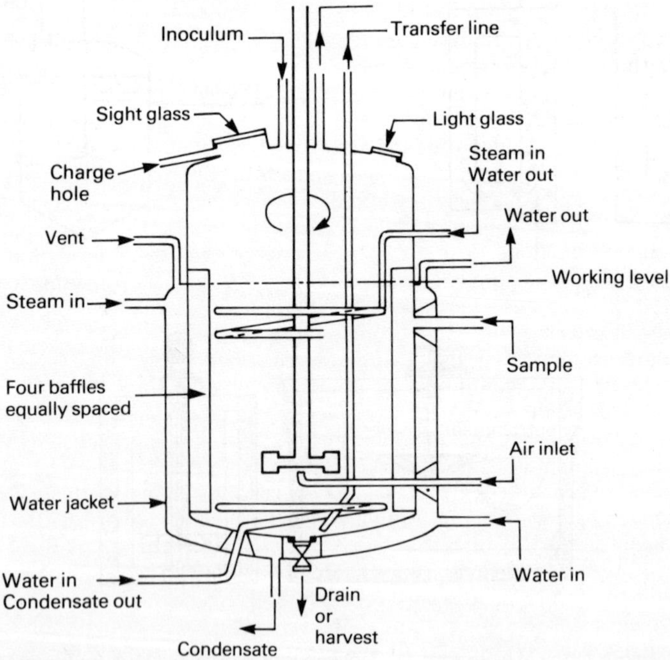

Fig. 13 An industrial continuous stirred tank reactor. Reproduced by permission from Atkinson (1974).

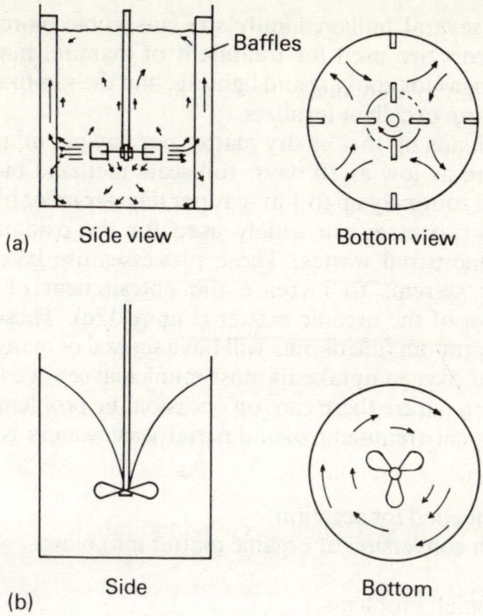

Fig. 14 Mixing patterns in bioreactors: (a) Radial flow pattern in baffled bioreactor, (b) Unbaffled flow pattern. Reproduced by permission from Oldshue (1983).

flow in a specified direction (Figure 12c). This has been achieved by incorporating draft or baffle tubes to give an 'internal recirculation' of the fluid or by 'external recirculation' using a recirculation pipe.

Large amounts of organic wastewaters from domestic and industrial sources are routinely treated by anaerobic and aerobic bioreactor systems. In the absence of free oxygen, certain specialized microorganisms are able to convert biodegradable organic material to methane, carbon dioxide and new microbial cells. Approximately 90% of the chemically bound energy in the original organic material is recovered as methane, 5 to 10% of the energy is used for formation of new microorganisms and about 3% wasted as heat. This is in sharp contrast to aerobic degradation processes where about 60% of the energy available is present in new cell growth and about 40% lost as process heat.

The most common type of anearobic bioreactor or digester is the CSTR (Figure 12d) operated in a continuous or semi-continuous manner. With this system concentrated wastewaters—for example, sludge from municipal sewage treatment—are mixed with the anaerobic microoganisms at approximately 30°C and the hydraulic retention time (the average time the water remains in the bioreactor) is selected to give efficient wastewater stabilization and high methane yield. For medium strength wastewaters from food and fermentation industries, techniques have been devised to retain the microbial biomass for longer periods in continuously operating systems. Thus solid retention time is uncoupled from the fluid retention time and high microbial concentrations can be obtained in the digesters thus giving high degradation rates. For very dilute wastewaters, for example, municipal sewage, very long solids retention time is required and this can only be achieved with fluidized bed processes (see Chapter 4).

The outstanding example of methane fermentation is the Chinese biogas programme where several million family size anaerobic biorectors have been built. Such bioreactors are used for treatment of manure, human excreta and straw, producing biogas for cooking and lighting, and the sanitization of the waste which then becomes an excellent fertilizer.

With volumetric loads up to 4 kg dry matter per day per m^3 of bioreactor and mean residence time as low as 10 days, full-scale methane bioreactors can be expected to produce routinely up to 1 m^3 gas per day per m^3 of bioreactor.

Activated sludge processes are widely used for the oxidative treatment of sewage and other industrial wastes. These processes use batch or continuous agitated bioreactor systems to increase the entrainment of air to optimize oxidative breakdown of the organic matter (Figure 12e). These bioreactors are very large and for optimum functioning will have several or many agitator units to facilitate mixing and oxygen uptake of most municipal sewage treatment plants. Because of their open nature there can, on occasion, be problems of smell.

Anaerobic biological treatment of industrial wastewaters is becoming more accepted because

(1) no energy is needed for aeration;
(2) there is a high conversion of organic matter into biogas, which can be used as fuel;
(3) there are no smell problems;
(4) little surplus sludge is produced;
(5) by manipulation of microbes, high value products can be formed.

Media design

Media are designed to meet the nutritional demands of the producer organism, the objectives of the process and the scale of the operation. For most large scale biotechnological processes cost, availability and handling properties of the medium components are major factors in determining selection.

The basic nutritional requirements of heterotrophic microorganisms (animal and plant cell requirements are considered later) are an energy or carbon source, an available nitrogen source, inorganic elements and, for some microorganisms, growth factors. For most biotechnological processes carbon and nitrogen sources are more often derived from relatively complex mixtures of cheap natural products or byproducts (Table 14), while trace metals are normally present in sufficient amounts in the tap water or in the main raw materials. Salts are often added as supplementary sources of nitrogen, phosphorus, sulphur or calcium. Growth factors when required may be supplied in pure form but for economic reasons would generally be added as a plant or animal extract. The main types of growth factors required are B-group vitamins or related compounds, certain amino acids and some fatty acids. Proper balance of carbon and nitrogen sources can be important to the pH pattern of processes if pH control is not applied. For most processes the nutrients have to be dissolved in water. In batch systems nutrients are normally all present in the initial volume. Fermentation reactions can be further regulated and controlled in batch cultures by feeding some of the nutrients on a specific rate basis (fed-batch culture). In this way essential inducer concentrations can be maintained. Nutrient availability will exert strong physiological control over fermentation reactions and product formation.

Table 14 Sources of carbohydrate and nitrogen for industrial media

Sources of carbohydrate		Sources of nitrogen (% nitrogen by weight).
Glucose	— Pure glucose mono hydrate, hydrolysed starch	Barley (1.5–2.0).
		Beet molasses (1.5–2.0).
Lactose	— Pure lactose, whey powder	Corn-steep liquor (4.5)
Starch	— Barley, groundnut meal, oat flour, rye flour, soya bean meal	Groundnut meal (8.0)
		Oat flour (1.5–2.0)
		Pharmamedia (8.0)
Sucrose	— Beet molasses, cane molasses, crude brown sugar, pure white sugar	Rye flour (1.5–2.0)
		Soya bean meal (8.0)
		Whey powder (4.5)

Adapted from Rhodes and Fletcher (1966).

Economy is of paramount consideration in medium formulation since raw material can account for 60 to 80% of the variable costs of fermentation processes. Raw material input to a fermentation will be largely dependent on the cost of the material at a particular time since commodity market prices fluctuate with seasonal and other variables. Choice of materials will also depend on handling and storage costs, ease of formulation and sterilization, while health and safety considerations must also be included.

Product optimization will be strongly influenced by medium formulation and specific nutrient availability during the fermentation. Thus, if biomass or a growth-associated product is the objective of the fermentation then the medium must permit maximum growth potential throughout. For a compound that is not growth limited, e.g. organic acids, antibiotics etc., a medium is developed which, after an initial growth phase, will become deficient in one or more nutrients. Depending on the nature of the process under study, particularly if a secondary metabolite is required, limitation of phosphorus, nitrogen, carbohydrates or trace metals have been successfully used. Some processes require the presence of an inducer in the medium while other processes may be repressed by a component of the medium. Catabolite repression is a particularly common problem with enzyme production and is manifested usually in media containing glucose. Repression can be avoided by replacing glucose by slowly fementable carbohydrates or partially hydrolysed starch. Incremental or continuous feeding of a concentrated component can also be used for specific fermentations.

The composition of an industrial medium is not only based on the requirements of the fermentation stage of the process but also on the subsequent purification steps. The formulation of the medium should also aim to produce a final fermentation broth that is low in viscosity, has an easily separable cell mass, and is low in residual compounds that could affect final product specifications.

Sterilization methods for media must achieve maximum kill of contaminating microorganisms with minimum temperature damage to medium components or the precipitation of minerals. Bacterial endospores can constitute a serious problem in medium stabilization since they are only killed at temperatures in excess of 100°C. At these temperatures many media components are labile and could be destroyed. In such cases alternative sterilization methods involving filtration or radiation must be applied.

For most media batch-wise sterilization in the bioreactor is still the method of choice, although continuous methods are gaining wider acceptability. Continuous sterilization processes with short holding time at temperatures in excess of 120°C can give an efficient spore kill without serious detrimental effects on medium nutrients. In practice continuous sterilization is obtained by passing the medium through a heat exchanger where it is raised to the required high temperature in a short time. The medium is then passed to a holding coil, maintained at this temperature for a predetermined time, and finally rapidly cooled by counter-circulation in the exchanger against cool input medium and then cold water. High-temperature/short-time processes result in improved preservation of growth factors and less development of colour. Recovery of heat may be an additional advantage. Direct steam injection (<140°C) can also be used for rapid sterilization.

Instrumentation and process control in bioreactors

All living organisms are subject to a wide range of intracellular and extracellular regulatory factors, such as temperature, pH and O_2. The identification and role of each of these factors has been derived from traditional physiological and biochemical investigations. In all biotechnological processes it is essential to optimize productivity. This can only be achieved by identifying and controlling the many factors that are known to regulate organism activity. The characteristic environmental parameters that control biological activity can be physical, chemical and biochemical and are not always easy to identify and measure (see Table 12).

Instrumentation of bioreactors has become increasingly important for measuring specific parameters, recording them, and then using this information to improve and optimize the process (Table 15). In practice, for process control a measurement is obtained by a *sensor* which can then be compared with a set point. A discrepancy between the two values is then used to alter the position of an *activator* that can manipulate the process, thus ensuring that the measured value is brought closer to the set point. The physical device that allows the measurement

Table 15 Instrumentation currently associated with bioreactors

Parameter measured	Equipment used
Pressure	Pneumatic or strain gauge sensor
Fermenter contents magnitude:	
Volume	Differential pressure cells at top and bottom
Weight	Fermenter mounted on load cell
pH	Glass and reference electrodes
Stirrer speed	Tachometer
Air flow	Variable area flowmeter, orifice plate
Carbon dioxide production	Infrared analyser
Oxygen consumption	Zirconia analyser
Dissolved oxygen	Polarographic oxygen electrodes
Temperature	Thermocouple, resistance thermometer, thermistor, capillary

From Atkinson and Mavituna (1983).

to be compared with the set point is a *controller* and in this way a control loop, able to regulate a specific factor, will consist of a sensor, controller and an activator.

Bioreactor control measurements can be made in either an on-line or an off-line manner. For on-line measurement the sensor is placed directly into the process stream whereas for off-line measurement a sample is removed aseptically from the process stream and analysed. Ideally a sensor should be steam-sterilizable, give a reliable continuous signal and operate on-line. It should be easy to calibrate, be robust and should not affect the process. The lack of efficient on-line sensors for measurement and control is a major bottleneck in the development of fermentation technology. The more common types of sensors for on-line measurement include those for temperature, pH, pressure, liquid and gas flow rates, CO_2 and O_2. Complete bioreactor processing is still severely limited by a lack of reliable instruments capable of on-line measurement of important variables such as DNA, RNA, enzymes and biomass. Off-line analysis is still essential for these compounds and since the results of these analyses are not normally available until several hours after sampling, they cannot be used for immediate control purposes.

A few of the well proven on-line systems will be examined briefly together with some of the projected future types of on-line measurement techniques.

Measurement techniques

Temperature Temperature can influence biological processes by kinetic effects on reaction rates and catalytic effects on the activity or stability of enzymes. There are many types of sensors available to monitor temperature in bioreactors which include thermocouples, resistance thermometers, thermistors and capillaries. They all function by producing an output signal that can be used in a control loop. Large volume bioreactors need several temperature sensors to ensure reasonable temperature distribution throughout the fermentation volume.

pH Growth of most organisms is sensitive to pH changes, each group of organisms having specific optimum pH values. The optimum value of pH for product formation is usually different from that for growth. pH is believed to exert its main effect on cell wall permeability and on the rates of reactions associated with the enzymes bound to the outer cell wall. pH sensors or ion-selective electrodes have been extensively used in bioreactors. A modern pH probe can withstand steam sterilization and some considerable degree of handling.

Dissolved oxygen Most organisms used in bioreactors need a constant supply of oxygen. Dissolved oxygen (DO) probes measure the amount of oxygen dissolved in the process liquid. The relationship between organism activity and dissolved oxygen is often the major determinant for the eventual success of most biotechnological processes. There are two kinds of DO probes, galvanic and polarographic. They do not differ significantly in design, each carrying two electrodes in a glass or stainless steel housing. The main problems with DO probes are that they have a slow response and are not very reliable at low DO concentration. There can also be considerable time drift and they do not stand up to repeated sterilizations.

The rate of oxygen uptake within a bioreactor can be obtained by measuring oxygen concentrations in the inlet and outlet gas streams using a paramagnetic oxygen analyser or a mass spectrometer which is faster but much more expensive.

Dissolved carbon dioxide Determination of dissolved CO_2 levels in a fermentation broth provides considerable information about a process. Membrane-covered electrodes are now available for dissolved CO_2 measurements but they are expensive and lack reliability. On-line measurement of gaseous CO_2 can be achieved with spectrophotometric, gas chromatographic or mass spectrometric methods.

Immobilized enzyme probes These new generations of probes are produced by immobilizing enzymes near the surface of an electrochemical sensor, usually a DO or pH probe. The enzyme can react with a specific molecule producing a response which can be detected by the electrochemical sensor. Enzyme electrode prototypes have been reported for many compounds including glucose, sucrose, urea, pyruvate, penicillin, ethanol, lactate, methane, cholesterol and several amino acids. These electrodes are highly specific and not affected by other compounds, are easy to calibrate and have high senstitivity. However, they have long response times, are not steam-sterilizable and are difficult as yet to use in on-line control loops. Yet, undoubtedly, their specificity and unlimited range of applications make enzyme probes an exciting area for future development in fermentation technology.

Biomass It is not yet possible to have automated on-line analysis of biomass at a realistic cost. Indirect methods have been considered based on material balances of gaseous oxygen and carbon dioxide. In practice, biomass determinations including protein, DNA and RNA must still be performed as a time consuming off-line activity.

Data processing The data gathered from various on-line and off-line measurements were originally stored in notebooks and recorder charts but now more often will be stored in the memory of a computer or magnetically encoded on tapes or disks. In a computer, data for a particular biotechnological process will be stored in the form of files which in practice will cover medium formulation, the fermentation process and the intricacies of downstream processing. The storing and organization of historic data is now a major operation in most industrial fermentation processes. The computer has become a valuable, indeed indispensible, means of understanding and controlling fermentation processes. Thus in one functional unit it is now possible to have process monitoring, sequence control, continuous control, calculation, data storage, data supply and reporting. Further applications of computers in biotechnology include nucleic acid and protein sequence analysis, restriction map generation, simulation of molecular processes and experiment planning and debugging (Friedland, 1983).

Mass and energy transfer

Growth and metabolism of organisms in bioreactor systems requires the continuous supply and uptake of essential nutrients together with the elimination of toxic waste metabolites. The exchange of materials between the surface of the organism and its external environment is known as mass transfer. Mass transfer and concomitant growth and biological activity of the organisms can only be achieved if certain forms of energy can be supplied to or removed from the whole bioreactor system. Thus heat energy is required for sterilization and process temperature control, and mechanical energy for stirring and mixing of bioreactor contents: as a result of biological activity, metabolic energy will be generated.

Mass and energy transfer processes represent a major part of all fermentation processes and furtherance of their understanding will be of major importance to maximizing biotechnological processes. The principles governing mass and energy transfer are derived from chemical engineering practice.

Mass transfer Central to all bioreactor processes will be the need to supply the correct amount of nutrient to meet organism or metabolic demand for optimum growth and/or product formation.

A nutrient component will need to cross through several resistances during passage from the source phase (e.g. oxygen sparger, nutrient feed inlet) into the organism phase. These phases may be continuous or dispersed. The continuous phase will normally be liquid (aqueous or non-aqueous) but can sometimes be gaseous, while the dispersed phase can be gaseous (air, CO_2, methane, etc), liquid (hydrocarbons) or solids (microbial pellets, flocs, immobilized cells or solid substrates). The locations of the mass transfer resistances are shown in Figure 15.

Most organisms used in biotechnology are aerobic and consequently oxygen mass transfer has always been an important consideration when designing bioreactor systems. The rates of mass transfer between gas and liquid phases will be strongly influenced by the solubility of the gaseous components in the liquid. Unfortunately oxygen is only sparingly soluble in water and this has meant developing oxygen dispersion techniques to maximize the availability of the oxygen to the organism in the bioreactor. Methods of monitoring dissolved oxygen in aqueous solution and in the gaseous phase have been well developed and are considered elsewhere. Measurement of $k_L a$ or dissolved oxygen concentrations are important for knowing the availability of oxygen to the organisms in a process.

Air or oxygen is introduced into a bioreactor system by way of an orifice producing bubbles and the size of the bubbles together with the velocity at which

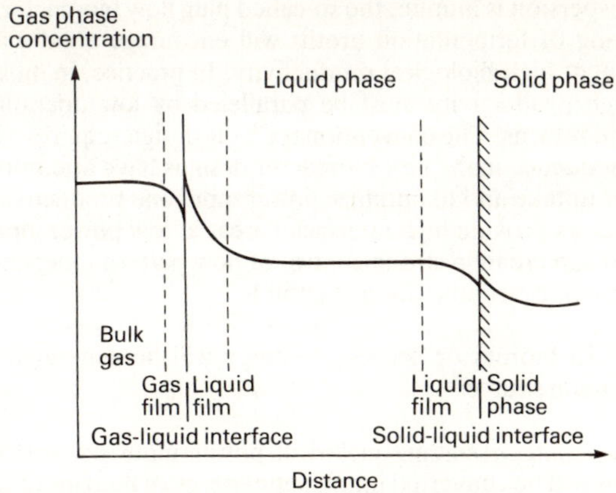

Fig. 15 Mass transfer resistances in a system with gas, liquid, and solid phases. Reproduced by permission from Sinclair and Mavituna (1983).

they rise in the fermentation liquid will determine the rate of oxygen transfer. In non-mechanical agitated systems, such as tower bioreactors, bubble diameter will determine the interfacial surface area and bubble rise velocity. Conversely, in mechanically agitated vessels (CSTR) bubble size will depend on the level of bulk turbulence and the physical properties of the liquid rather than the orifice or sparger design. Excessive viscosity of the liquid will further adversely influence oxygen transfer.

Mass transfer across solid/liquid phase boundaries in fermentation systems are not strongly influenced by the hydrodynamics of the whole system. All transfers will be by molecular diffusion and the rate of diffusion of nutrients (including oxygen) to the cells and of metabolites from the cells will limit the rate of biological activity. Diffusion will always be accompanied by reaction, i.e. either consumption or production of the diffusion molecules.

Aerobic fermentation can be seen as a gas/liquid mass transfer step, a liquid/solid mass transfer step and a chemical or biochemical reaction step. Mixing has a greater effect on the gas/liquid mass transfer step than on the liquid/solid step.

The complete and uniform dispersion or mixing of the medium components within a bioreactor becomes increasingly more complex with increasing volume of the bioreactor. The mechanism of dispersion throughout the bioreactor volume will involve the bulk flow of liquid together with mixing by entrainment via the turbulent eddies that are generated either by mechanical agitation or the turbulent flow of the liquid. Mixing will be particularly important for the addition of air and of nutrients in fed-batch or continuous culture.

Mixing mechanisms should aim to optimize both axial and radial dispersion thus ensuring complete preservation of the fermentation components (medium and organism) in all parts of the bioreactor. Seldom is this completely achieved and in large-scale fermentation rapid and near uniform dispersion of added ingredients can only be achieved by using many ports of entry into the bioreactor. Indeed, the large ICI Pruteen bioreactor utilizing methanol as substrate involves 2600 individual entry points. When broths are non-Newtonian and highly viscous there will always be parts of the bioreactor volume that are stagnant and not fully participating in the overall fermentation process. When axial dispersion is near zero and radial dispersion is infinite, the so-called plug flow bioreactor occurs.

Thorough mixing of fermentation broths will encourage good mass transfer rates and subsequent high biological productivity. In practice, in most industrial fermentations high productivity must be paralleled by low operating costs to maximize financial returns. The conventional CSTR design requires a high power input and in consequence many new bioreactor designs have attempted to lower the specific power uptake and to optimize power input and productivity. Some of these new bioreactors provide high interfacial areas at low power inputs (bubble column, etc.) and can create a strong control of flow pattern independent of gas hold up by an inserted draft tube (loop reactors).

Energy transfer In bioreactor processes energy will be generated within the system from four main sources.

(1) *Mechanical agitation energy*: electrical power input or work to stirring mechanisms will be converted into kinetic energy of fluid motion which will be dissipated in various ways to appear eventually as heat.
(2) *Gassing and aeration energy*: some energy will be dissipated at the sparger

holes and in creating turbulent eddies although most of this type of energy will represent the work done by the gas as it expands when moving through the liquid static head.

(3) *Metabolic energy*: the organisms within the bioreactor oxidize organic molecules and some of the energy will be dissipated as heat.

(4) *Enthalpies*: heat can be generated in the bioreactor if the input streams are at a higher temperature than the bioreactor contents.

Build-up of heat in a bioreactor can result in the temperature of the fermentation broth exceeding the optimum temperature for productivity. Consequently heat may have to be removed from a bioreactor. Heat can be removed across solid boundaries to the surrounding air or to cooling water in internal coils, external jackets or external heat exchangers. Examples of cooling systems for bioreactors are shown in Figure 16. When bioreactors are run at temperatures above ambient then, even allowing for the generation of heat by the above mechanisms, a source of heat may be required to maintain an optimum temperature.

Scale-up

Most biotechnological processes will have been developed at laboratory scale and commercial success will largely depend on the ability to scale-up the process first to pilot plant level and then to commercial scale. The achievement of successful process scale-up must fit within a variety of physical and economic constraints. To date it has not been possible to establish a uniform design procedure or a simple

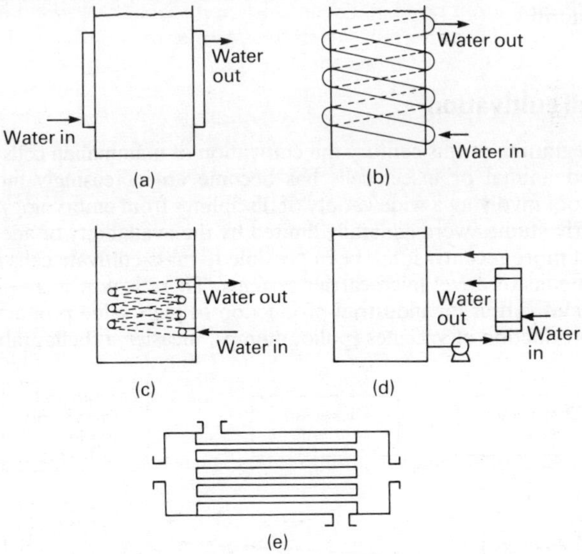

Fig. 16 Methods of heat removal. (a) Jacketed vessel, (b) External half coils (exaggerated dimensions), (c) Internal coil, (d) External heat exchanger, (e) Shell and tube heat exchanger (schematic). Reproduced by permission from Sinclair and Mavituna (1983).

method for process scale-up. For most products of commercial interest the producer organism (microbe, animal cell etc.) will optimize production under limitation of a nutrient or due to unfavourable environmental conditions. Microbial culture systems are usually heterogeneous and process problems will occur primarily from the transfer of nutrients and to a lesser extent from the transfer of products. (Fig. 17; Brown 1982).

The identification of some of the controlling factors can usually be made with laboratory scale bioreactors (5 to 101) while others may require pilot scale bioreactors. During the change from one scale of operation to another some aspects will remain constant (Table 16); others will change as a direct result of the increased scale (Table 16) while yet others can be within the control of the process engineer.

A pilot plant is in reality a large-scale laboratory which has been designed to give flexibility for equipment accommodation (bioreactors, pumps, heat exchangers, storage facilities, electrical and piping services, etc.), and adaptability for process operation. Engineering design criteria will incorporate simplicity, economy of operation and containment of air and water wastes. Each pilot plant will have its own particular features and design requirements.

Pilot plant bioreactors range from 100 l to 10 000 l total volume and the larger pilot bioreactors can on occasion be used as production units. Full scale industrial bioreactors can range between 40 000 and 400 000 l in volume.

When moving from the laboratory to the industrial scale major problems arise with heat removal, oxygen transfer and availability and insufficient diffusion of the components in the medium, for example nutrients, toxic metabolites and acids or bases. The management of scale-up problems requires high capital investment in mixing and aeration, in monitoring and control devices, and in stringent maintenance of sterility.

Animal cell cultivation

From the beginning of this century the cultivation of mammalian cells as well as cold-blooded animal or insect cells has become an increasingly more useful laboratory tool involving a wide variety of disciplines from embryology to microbiology. Early studies were generally limited by the availability of adequate cell numbers but more recently it has been possible to mass-cultivate cells in suspension or by means of novel microcarrier systems. There is now increasing use of animal cell cultivation for industrial production of high value products, in particular for production of vaccines (polio, mumps, measles, rubella, rabies, etc.),

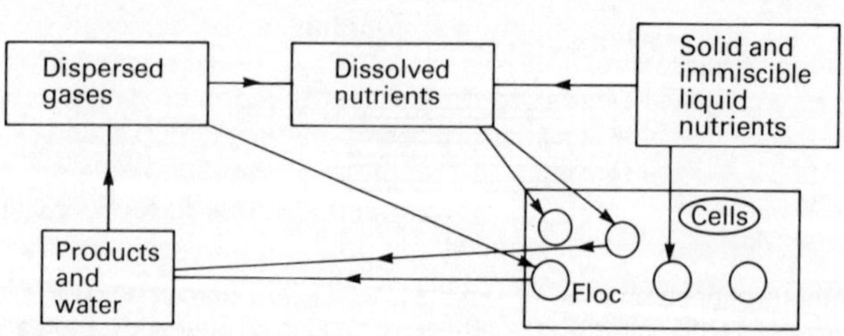

Fig. 17 The heterogeneous nature of microbial culture systems.

Table 16 Factors associated with scale-up

Items which remain constant	Items which change	Items which are in control of operator
(a) Basic design of the bioreactor and pipework arrangements	Bioreactor dimensions will increase	Components in the medium might be changed in concentration and also obtained in cruder, commercial grades
(b) Number of personnel required to operate the bioreactor	Mass of material and of equipment will increase	Overall pressure on the fermenter is often increased above atmospheric pressure
(c) Species and strain of microorganism	Quantity of material to be handled before and after the culture process will increase	Energy input per unit mass will, if possible, be reduced
(d) Medium components and special nutrient limitations	Number of vessels to process a given quantity of materials will decrease	Impeller speeds will be reduced accompanied by increased fluid flow and velocities but perhaps by decreased turbulence
(e) Sterilizing and operating temperatures	Cost of infections and other failures in the process operation will increase	Circulation times and mixing times will be adjusted to maintain adequate blending of the additions to the culture
(f) Rheological properties of the culture	Overall sterilizing cycle will change due to increased heating and cooling times	
(g) Bubble sizes in range 0.005 to 0.2 mm diameter	Area-to-volume ratio will decrease and wall effects will virtually disappear	
(h)	Surface aeration will decrease substantially	

D. Brown, Department of Chemical Engineering, UMIST, Manchester.

interferons, hormones and immunological agents including insulin, monospecific antibody molecules derived from hybridoma cells and clotting factors such as plasminogen and plasminogen activator. Although many of these proteinaceous materials may soon be obtained by cloning the necessary genes into simple microbial cells such as *E. coli* or *S. cerevisiae* many will be too complicated to warrant this approach and supply will continue to be achieved by mass cell cultivation techniques, similar in many ways to traditional fermentation practices previously described. The main problems that are encountered in mass cultivation of animal cells include high sensitivity of cells to impurities in the water supply, the cost and quality control of media and the need to avoid contamination by more rapidly growing microorganisms. Animal cell cultures have been significantly improved or altered by techniques of cell fusion (to form hybridomas), chromosome transfer and engineered plasmid infections (Chapter 2).

The Cell Line Cells which have been derived from animals and induced to proliferate in artificial culture are designated *in vitro* cultures and exist as primary cultures, secondary cultures or cell lines. Primary cell cultures are cells derived directly from the particular organ or tissue of the original organism or from blood and represent morphologically and functionally non-uniform populations. Free cells can be obtained from tissue pieces by repeated trypsinization giving cells in suspension or establishing monolayer cultures attached to surfaces such as glass, collagen or metal. Secondary cells will be obtained following *in vitro* passage of primary cells. Generally growth of secondary cells will be limited and will eventually cease. However, some cells will continue to grow and will establish a continuous cell line, which should be capable of being passaged indefinitely *in vitro*. A cell line must demonstrate the ability to be passaged at least 70 times from primary isolation before being designated continuous. Some cell lines have been derived from tumours while others have come from normal cells. Extensive cell culture lines have now been established on a worldwide basis and, in particular, the American Type Culture Collection holds over 500 strains.

Culture medium The selection of culture medium will be dependent on the specific nature of the cell line to be used. Culture medium will have four fundamental components: base medium, serum, additives, and buffering system. Each has a specific function in the final formulation and relative proportions must be determined from previous studies combined with the specific needs of the experiment. Although media may be prepared within the laboratory they can be purchased either as a liquid, 10 × liquid concentrate, powder, or autoclavable powder. An important feature of all cell cultivations is the purity of the water supply. Cell growth inhibitors are regularly found in water used to make up the medium. Reverse osmotically purified water has generally been shown to be the most suitable source for media preparation. Reconstituted medium is preferably sterilized by membrane filtration. Serum has long been an essential component in media design. However, since each batch of serum will be different in its ability to support cell growth it is often best to bulk several batches, redistribute, and store it. Serum can also be in short supply. The precise function of serum in the complete cell culture medium is still not fully understood, but it is considered to have at least four roles. These are to provide a carrier of low molecular weight nutrients; hormones (insulin) essential for cells in culture, protection against mechanical damage from sparging, and to stimulate the adherance of anchorage-dependent cells on surfaces. Sera vary from animal to animal and with age, sex and nutritional state of the donor.

Because of these problems there has been development of serum-free media or media with less costly and more readily available types of serum. More chemically defined media producing more reproducible growth properties are now available although most still require the addition of expensive growth factors or hormones. Other additives include antibiotics and all must be added using aseptic techniques. Buffering systems will normally be a synthetic mixture of inorganic salts which function to maintain the pH and osmotic pressure of the medium. Growth of animal cells in a nutritionally complete tissue culture medium is usually optimum when the medium is buffered at a pH in the range 7.2–7.4.

Microbial contamination at all levels of cultivation is an ever-present hazard of animal cell cultivation. The rich media used in cultivation of animal cells are abundantly suitable for most common microbial contaminants. Laminar airflow cabinets offer special protection for many aspects of culture handling and larger scale cultivation must strictly adhere to aseptic operation procedures.

Cell cultivation systems Although the mass cultivation of animal cells is more complex and demanding than that of microbial cells, the technology of the various processes is rapidly approaching the levels achieved with most microbial systems. There are two main types of cultivation system used for producing animal cells in mass culture—monolayer and suspension cultures (Figure 18). Monolayer cultivation systems require a solid support on which the cells may grow, a factor which makes monolayer culture more difficult to scale up. In contrast suspension culture systems grow freely throughout the bulk of the culture behaving in a manner similar to most microbial systems. Scale up of suspension culture systems presents few problems and therefore offers more advantages for use in large scale commercial processes.

From a historical viewpoint the development and use of viral vaccines has led to major achievements in medicine and biotechnology. The oldest vaccines, such as smallpox, yellow fever and rabies, were propagated in animals while later it was found that the embryonated hen's egg could serve as an excellent substrate for several viruses. With the development of techniques for monolayer tissue cultures of animal and human cells the production of vaccines has dramatically changed. In this way many different viruses could be propagated under defined controlled conditions *in vitro* to produce, for example, vaccines against poliomyelitis. In recent years recombinant DNA technology has allowed the synthesis of selected virus structural components in bacteria while monoclonal antibody techniques are now able to define the antigenic epitopes on the surface of the virions, the particles primarily responsible for the induction of immunity against disease.

Monolayer or anchorage-dependent culture For most animal and human cells

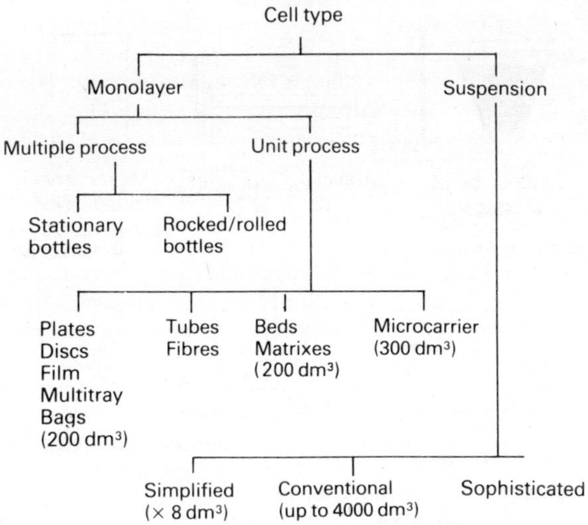

Fig. 18 Equipment used for large-scale production of animal cells. Reproduced by permission from Spier (1982).

successful cultivation requires the presence of surfaces to which the cells can become anchored. The continued growth of these anchorage-dependent cells is contingent upon the surface available and when this has been completely covered or closed over with a monolayer of cells growth will cease. This is known as 'contact' or 'density' inhibition and has represented the most serious impediment for large scale operation of this cell type. Most cell types of biotechnological potential, namely those suitable for producing viral vaccines, interferons and hormones, exhibit this form of growth.

A wide range of methods suitable for anchorage-dependent cells have been described and include various forms of roller bottle systems, plate and packed-bed bioreactors, and several forms of hollow fibre surfaces (Figure 19). For most of these techniques it is necessary regularly to perfuse the cells with fresh medium and try to ensure uniformity of parameters such as pH and CO_2. Growth and yield of cells and products are less than theoretically expected and the whole system is difficult to scale up because of the surface-to-volume ratios required.

Most of these disadvantages can now be overcome by cultivating the anchorage-dependent cells on various forms of small particles (microcarrier systems) which can be held in suspension with gentle agitation. In this way cells can now be grown in a more homogeneous culture system with greater potential for control and scale-up. The number of microcarriers per cm^3 can range from 5000 to 25 000

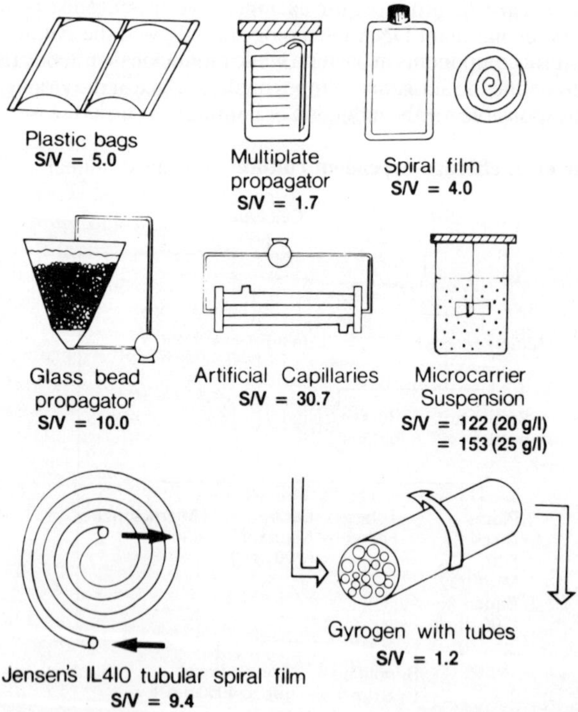

Fig. 19 Surface-to-volume ratios (cm^{-1}) of various methods used to culture anchorage-dependent cells.From Glacken *et al.*, 1983.

offering a massive surface area for colonization together with a particle size easily held in suspension. Initially collodion-coated DEAE–Sephadex A50 (Pharmacia) beads were used for microcarriers but now serum-coated beads or heat-treated foetal-calf-serum-coated beads arc used. Other forms include treated polystyrene and polypropylene or diethylaminoethyl-substituted cellulose fibres. Very high surface-to-volume ratios can be achieved with these microcarrier systems, for example 1 mg of Cytodex microcarrier beads per cm^3 of culture fluid creates a surface area of 5 cm^2 cm^{-3} which in turn can support 750–1000 × 10^3 cells cm^{-3}. The concentration of beads may be increased up to 5–7 mg cm^{-3} without toxic effects.

Standard bioreactor approaches may be used for microcarrier cultivation although it is advisable to use low stirrer speeds to avoid shear effects and cell detachment, but round-bottomed fermenters are preferred to avoid settling and accumulation of the carriers. The cultivation systems allow easy medium changes or washing of the cell cultures. Continuous operation is also possible by using special filter attachments that retain the carriers in solution and allow continuous medium change. The pH can be controlled by increasing the CO_2 concentration in the gas mixture or by adding $NaHCO_3$ or NaOH. The introduction of 0_2 into the bioreactors is difficult and few methods use submerged sparging since this creates foaming problems. Instead 0_2 or air is blown over the surface of the culture and slow oxygen exchange is achieved. Cells may be removed from microcarriers by trypsinization or by physical manipulations.

Suspension cell culture Suspension cell systems have been developed for several cell types including human cells. The obvious advantages of these systems include larger culture volumes and reduced costs, cells can achieve a steady state as they are exposed to more uniform medium, and scale up is much easier.

Transformed human cells which have many properties similar to tumour cells can be grown easily in suspension cultures. Concern has been expressed about the use of transformed cells for production of pharmaceutical-type products because of a possibility of transfer of some harmful transmissable agents into the product. High standards of product purification must be adhered to rigorously, to exclude this possbility. Transformed cells such as lymphoblastoid cells are being used to produce large amounts of human interferons.

The basic fermentation technology of suspension culture primarily follows the normal principles of microbial cultivation described previously. The interferon process uses stainless steel tanks which are magnetically stirred and with volumes up to 4000 l.

Aeration of large suspension cultures can present obstacles to good cell growth. High intensity bubbling to achieve maximum bubble dispersion can cause physical damage to cells and, for particularly fragile cells systems, pure oxygen diffusion has been used.

Although the technology of mass cell culture is rapidly evolving there are still many problems to be overcome before these systems truly match the well developed microbial processes.

Plant cell cultivation

Horticultural practice has long made use of vegetative cloning of plant parts (shoots, leaves, etc.) for selective and rapid propagation of many plant species. Subsequently, plant tissues such as root tips, cambium or plant tumours were found to be relatively easy to grow aseptically in artificial cultures. With careful selection of medium and other environmental factors it was possible with many species to achieve differentiation and eventual regeneration of whole plants. The four stages of the complete cycle are:

(1) establishment in culture;
(2) organogenesis;
(3) plantlet amplification;
(4) establishment in soil.

The major commercial use of this type of tissue culture has been towards the rapid propagation of plants both for stock purposes and for sale.

This ability of the horticulturist to clone large quantities of identical plants from a single parent in a short period of time highlights the regenerative ability of many plant tissues. This practice is now widely used with many commercial advantages including continuous production throughout the year, rapid production of new varieties obtained by conventional genetics, virus-free stock production from shoot tip cultures and genetic improvements of plants using haploids. Furthermore, this technique could be of immense value for eventual mass production of genetically modified cells derived from protoplast fusion or recombinant DNA technologies. Tissue culture can also be used in the long-term storage of specialized germ plasm, providing stable genetic material, reducing storage space and decreasing maintenance costs.

Plant cells can also be cultured in various ways to produce an array of valuable pharmaceuticals, dyes and flavours that are normally derived from the whole plant. Plants still produce about 25% of prescribed medicines and together with the many useful fine chemicals, plant cell culture could offer a suitable means for producing these low volume/high value products by procedures similar to those used in traditional microbially-based fermentations. Examples of these products which have been produced in tissue cultures include steroids, cardiac glycosides, antitumour alkaloids, codeine and 3,4-dihydroxyphenylalanine. Furthermore, there is every expectation that in the future this form of biotechnology will be a major source of antibiotics, insecticides, food flavouring and colouring agents, perfumes and emulsifiers. However, before a plant chemical product can be produced by this biotechnological route two important questions must be considered: is there sufficient commercial value to warrant the investment, and will the biotechnological route overcome problems inherent in conventional production.

The application of plant cell cultures to biotechnology will require the establishment of mass cell culture techniques. Callus cultures of many plants can now routinely be achieved and maintained on suitable medium where the cells grow in a disorganized mass of mainly undifferentiated cells. When grown in liquid suspension culture, plant cells will grow as a mixture of single cells and clumps of cells. Uniformity of culture is a well sought after goal in plant cell cultivation and well dispersed cultures can be achieved by mechanical, chemical or enzymic destruction of cell clumps. Plant cells are not attachment dependent.

Complex media using coconut milk and other undefined growth regulators were necessary for early plant cell studies but nowadays defined media are widely in use. Defined media generally contain a mixture of inorganic salts and a carbon or energy source, normally glucose or sucrose, together with specific vitamins and growth regulators such as indole acetic acid, 2,4-dicholorophenoxyacetic acid or cytokinins. Variations in media are related to the type of cells being cultured and to the products required. Media used for plant cell culture are both less complex and less expensive than those used for culturing mammalian cells.

The methodology of mass cultivation of plant cells in liquid, submerged culture has been derived largely from microbial methods. While the culture of plant cells is more complex than that of microorganisms (because of relatively low growth rates and the tendency of the plant cells to develop small tissue masses) the main aspects of heterotrophic nutrition, mass transfer, shear forces and fermentation process control are very similar to the microbial processes previously discussed in this chapter.

Doubling times of 20–40 hours are characteristics for most cell cultures (cf. bacterial doubling time about 30 minutes) and are related to the large size of plant cells. The volume of an average cultured plant cell may be up to 200 000 times that of a bacterial cell. The growth cycle can exhibit a lag phase followed by a phase only approximating to exponential growth and finally a stationary phase. The fragile nature of plant cells and their large size generate considerable practical problems for mixing, aeration and mass transfer. Various mixing systems have been considered for plant cell culture including flat blade turbines, propellers and bubble columns. Since plant cells show considerable tensile strength but low shear tolerance air lift bioreactors have been preferred to turbine stirrers. Air-lift bioreactors are also relatively inexpensive, mechanically simple and easy to sterilize.

The main area for commercial exploitation of mass plant cell culture will be with plant products of complex structures and stereochemistry produced naturally from plants of limited availability. Furthermore genes for the synthesis of such compounds are unlikely ever to be cloned into microorganisms because most plant secondary products are under multiple gene control. Suspension cultures have been found in many cases to give similar or even greater yields of products when compared with the parent plant, e.g. nicotine, alkaloids and ginseng. Profitable fermentation processes can only be expected with high value products (Table 17). It is unlikely that plant cell culture production of chemicals will be commercially attractive in most cases for at least another 20 to 30 years. However, the first

Table 17 Markets for some plant secondary metabolites

Compound	Use	Wholesale price	Estimated retail market
Vinblastine/ vincristine	Leukemia	\$5000/g	\$18–20M (US)
Ajmalicine	Circulatory problems	\$1500/kg	\$5.25M (World)
Digitalis	Heart disorders	\$3000/kg	\$20–55M (US)
Quinine	Malaria; flavour	\$100/kg	\$5–10M (US)
Codeine	Sedative	\$650/kg	\$50M (US)
Jasmine	Fragrance	\$300/kg	\$0.5M (US)
Pyrethrins	Insecticide	\$300/kg	\$20M (US)
Spearmint	Flavour; fragrance	\$30/kg	\$85–90M (World)

From Curtin (1983).

marketable chemical compounds produced from plant tissue culture—Shikonin, a dye and pharmaceutical—is now being sold in Japan at approximately $4000/kg.

Solid substrate fermentation

Solid substrate fermentations are concerned with the growth of microorganisms on solid materials in the absence or near absence of free water (Table 18). The upper level of solid substrate fermentation (i.e. before free water becomes apparent) is a function of absorbancy and thus moisture content which will, in turn, be dependent upon type of substrate. Biological activity ceases when the moisture content of a substrate is about 12% and as this value is approached microbiological activity becomes increasingly retarded. Solid substrate fermentations are not concerned with the fermentation of slurries (i.e. liquids with high levels of insoluble solids) or with the fermentation of solid substrates in a liquid medium. The most common substrates used in solid substrate fermentations are cereal grains, legume seeds, wheat bran, lignocellulosic materials such as wood and straw, and a variety of other plant and animal matter. The compounds are invariably polymeric molecules, insoluble or sparingly soluble in water, but are cheap, easily obtainable and represent a concentrated source of nutrients.

Solid substrate fermentations have great antiquity, having been practised in the Orient for hundreds of years. Many oriental food fermentations such as soy sauce, miso, tempeh, etc., have an essential solid substrate phase, while other solid substrate processes have been used to produce enzymes and chemicals such as citric acid. In the western hemisphere solid substrate fermentations have centred around composting of plant and animal wastes, ensiling, mushroom cultivation, and cheese manufacturing. Solid substrate fermentations of lignocellulose could well be a major industry of the future, producing biomass, ethanol, methane and many other commercially valuable products. Most microbial-based biotechnological products could be produced by solid substrate fermentations. The determining factor for such practice will depend on the relative economics when compared with liquid fermentation processes.

Table 18 Some examples of solid-substrate fermentations

Example	Substrate	Microorganism(s) involved
Mushroom production European and Oriental)	Straw, manure	*Agaricus bisporus* *Lentinus edodes* *Volvariella volvaceae*
Oriental fermentations	Wheat and soybeans	*Aspergillus oryzae*
Soy sauce	Soybeans	*Rhizopus* sp.
Tempeh	Peanut press cake	*Neurospora sitophila*
Ontjom		
Cheeses	Milk curd	*Penicillim roquefortii*
Leaching of metals	Low grade ores	*Thiobacillus* sp.
Organic acids	Cane sugar, molasses	*Aspergillus niger*
Enzymes	Wheat bran, etc	*Aspergillus niger*
Composting	Mixed organic material	Fungi, bacteria, actinomycetes
Sewage treatment	Components of sewage	Bacteria, fungi and protozoa

The types of microorganisms that grow well under conditions of solid substrate fermentations are determined in large part by the water activity factor (a_w). The a_w of the substrate quantitatively expresses the water requirement for microbial activity.

$$\ln a_w = \frac{-vm\phi}{55.5}$$

where v = the number of ions formed, m = the molar concentration of solute, ϕ = the molar osmotic coefficient, and 55.5 = the molar concentration of an aqueous solution of pure water.

Pure water has an a_w = 1.00 and a_w will decrease with the addition of solutes. Bacteria are mainly found at the higher a_w values while certain filamentous fungi and a few yeasts can grow at values between 0.6–0.7. Organisms that can tolerate and propagate at low a_w levels are in general the main organisms that occur in solid substrate fermentations.

Microbial types Solid substrate fermentations can occur in many different forms depending on whether the microorganisms utilized are indigenous, pure cultures or mixed cultures.

Fermentations that use indigenous microflora are primarily *ensiling* and *composting*. Ensiling is an anaerobic process involving agricultural vegetation and is carried out at 25–30°C for 1–2 weeks. *Lactobacillus bulgaricus* becomes the dominant organism, producing lactic acid and subsequently inhibiting potential putrefactive bacteria, and because of the absence of oxygen aerobic mould fungi cannot grow. Moisture level is critical at 50–65%, thus ensuring that the osmotolerant lactobacilli are active and dominant. In contrast, composting involves a succession of microorganisms from mesophilic bacteria, yeasts and mould fungi, through to thermophilic actinomycetes and fungi. Heat generation by biological activity is a serious problem and the compost must be turned mechanically to avoid sterilization. Composting for mushroom production is one of the most successful means of creatively using lignocellulosic materials.

Solid substrate fermentations using pure fungal cultures are best illustrated with the ancient Koji process for the fermentation of grains and soya beans with the fungus *Aspergillus oryzae*. The cooked substrate is inoculated with a pure culture of *A. oryzae* and grown in shallow layers in trays or in special rotating bioreactors to produce amylases and proteases to break down polymeric materials in the substrate. The Koji process is the basis for other types of fermentations including commercial enzyme production, organic acids and ethanol. The Koji process is also finding novel uses for biomass production from starchy materials (Raimbault/Alazard process) and from cellulosic materials (Waterloo process).

Certain solid substrate fermentations deliberately use known mixed culture inoculations to achieve optimum end product formation. Thus straw can be more efficiently converted to fungal biomass using mixed cultures of *Chaetomium cellulolyticum* and *Candida lipolytica* than by using either fungus alone. Such fermentations are considered to have considerable potential for low technology conversion of lignocellulosic materials.

A feature characteristic of many solid substrate fermentations is the need to pretreat the substrate raw materials to enhance the availability of the nutrients, or to reduce the particle size to optimize the physical parameters of the fermentation.

Polymeric molecules may need to be partially hydrolysed to facilitate organism growth while suitable particle size can be achieved by numerous forms of physical treatment, for example ball milling. Cost aspects of pretreatment must be balanced with product value.

The design of solid substrate fermentation processes will be further controlled by the need to establish good mass and heat transfer characteristics. Inter-particle mass transfer and intra-particle diffusion are the two main mass transfer steps limiting solid substrate fermentations.

Inter-particle mass transfer In a solid substrate fermentation, particle size determines the amount of space within the substrate mass that can be occupied by air (the void space). Most fermentations involve aerobic microorganisms and transport of oxygen into the void space is a critical parameter controlling successful development of growth and product formation. Oxygen transfer into the void space is closely related to moisture level since high free moisture level effected by forcing air out will ensure low void space. In practice, satisfactory inter-particle transfer of oxygen can be achieved by careful mixing and aeration. Oxygen levels can be monitored within the substrate mass and when necessary intermittent mixing and aeration can be practised. It is also essential to avoid CO_2 build-up in the void spaces.

Intra-particle mass transfer Intra-particle mass transfer is concerned with the transfer of nutrients and enzymes within the fermentation substrates. In solid substrate fermentations using filamentous fungi, hyphae will grow over the particle surface as well as penetrating into the substrate mass. The aerobic hyphae will require the diffusion of oxygen to support their continued growth. The kinetics of oxygen diffusion in these complex solid substrate matrices is far from being understood.

Intra-particle mass transfer is also concerned with the role of enzymes in hydrolysing the water-insoluble polymers into soluble substrates to be utilized by growing fungi. These hydrolytic enzymes therefore have a major role in the overall degradation or utilization of the substrates. Substrates with open, porous structures are more easily degraded than those with less porous surfaces. The enzymic hydrolysis of cellulosic materials has been extensively studied with solid substrate systems and has been shown to be dependent on the action of an enzyme complex, cellulase, which involves three types of enzymic reactions:

(1) endo–β–1,4–glucanase which hydrolyses cellulose randomly producing glucose and cellobiose as end products;
(2) exo–β–1,4–glucanase which attacks the non-reducing end of a cellulose polymer chain and produces cellobiose;
(3) cellobiase or β-glucosidase which acts mainly on the cellobiose to form glucose.

Heat transfer Because of the high substrate concentrations per unit volume, microbial heat generation is much greater per unit volume than for liquid fermentations. Also, the low moisture content of these fermentations creates difficult conditions for heat transfer, so the control of temperature is more difficult than for liquid fermentations. In particular, heat removal from large-scale operations is a major problem. Heat removal can be facilitated by increasing the aeration rate or the frequency of aeration for systems that are not continuously aerated. A corollary to this is that when a fermentation temperature is too low then reduced aeration can generally result in increasing the operating temperature without the need to add an external source of heat. In practice, metabolic heat generation can be regulated by several means including adjusting the temperature and relative humidity of the air, the aeration rate and the agitation speed.

Bioreactors Solid substrate fermentations can be classified into fermentations without agitation, fermentations with occasional agitation and fermentations with continuous agitation. Fermentations without aeration include ensiling and a limited range of composting processes and will not be further considered. The latter two types of fermentation can be further subdivided into:

(1) fermentations with occasional agitation, without forced aeration (Figure 20);
(2) fermentations with slow continuous agitation, without forced aeration (Figure 20);
(3) fermentations with occasional agitation with forced aeration (Figure 20).

Group (3) above represents the main direction of bioreactor design in Japan and current systems provide a good environment for growth and product formation, controlled mechanical agitation, reasonable process control and sterile operation conditions. Most systems are designed for batch operation but several may be adapted for semi-continuous or continuous operation.

Process control in liquid fermentations has allowed increased monitoring of biological activity within the bioreactor and sophisticated computer-controlled feedback loops can be operated, giving better dynamic control of the process. In contrast solid substrate fermentations lag far behind in this respect. There is an acute lack of sensors: most existing sensors are of little value because they have been designed for the measurement of dissolved components or for physical characteristics of liquids.

The relative advantages and disadvantages of solid substrate fermentations compared with liquid fermentations are shown in Table 19. Future areas of improvement in solid substrate fermentations will be in substrate pretreatment, process control and fermenter design.

Summary

Growth is the increase of cell material expressed in terms of mass or cell number. The doubling time of an organism population is the time required for the doubling in biomass while generation time relates to the period necessary for the doubling of cell numbers. Mathematical equations have been developed describing the essential features of microbial growth in bioreactors. In batch cultivation, there is a continuously changing nutrient environment which is reflected in the physiological state of the culture. In contrast, in continuous culture, with controlled addition and removal of medium, a steady state of population characteristics can be achieved.

In fermentation technology, large numbers of cells are grown under defined, controlled conditions for biomass or product formation. In general the processes are carried out in a containment system or bioreactor, the main function of which

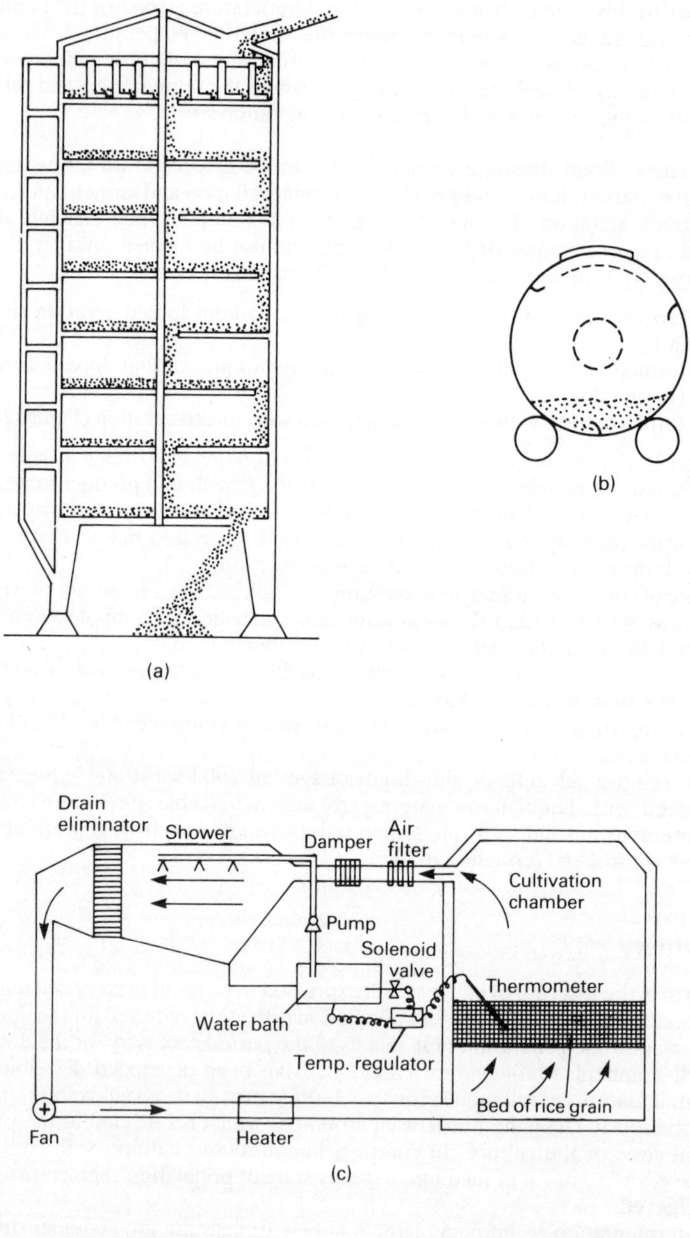

Fig. 20 Bioreactor systems for solid substrate fermentations. (a) Earp-Thomas tower bioreactor, (b) Rotating drum bioreactor, (c) Japanese Koji bioreactor.

Table 19 Comparison of solid substrate fermentations with liquid fermentations

Advantages	Disadvantages
Simple media with cheaper natural rather than costly synthetic components	Processes limited mainly to moulds that tolerate low moisture levels
Low moisture content of materials gives economy of bioreactor space, low liquid effluent treatment, less microbial contamination, often no need to sterilize, easier downstream processing	Metabolite heat production in large-scale operation creates problems
Aeration requirements can be met by simple gas diffusion or by aerating intermittently rather than continuously	Process monitoring, e.g. moisture levels, biomass, O_2 and CO_2 levels is difficult to achieve accurately
Yields of products can be as high	Bioreactor design not well developed
Low energy expenditure compared with stirred tank bioreactors	Product limitations
	Slower growth rates of microorganisms

is to minimize the cost of producing a product or service. Processes can be considered as conversion cost-intensive or recovery cost-intensive. Bioreactors can operate on a batch, fed-batch or continuous basis. The biocatalysts (micro-organisms, plant or animal cells or enzymes) can function in a free form or be immobilized. Optimization of a bioreactor process will involve minimizing raw materials and energy use and maximizing product formation.

Bioreactor design will depend on the nature of the process. The continuous stirred tank system (CSTR) is the most widely used bioreactor and can be operated for aerobic processes as well as for anaerobic processes. Mixing in a CSTR bioreactor is achieved by a mechanically operated central shaft equipped with blades or impellers. In contrast, tower bioreactors do not have mechanical agitation and mixing of contents is achieved by the rising air bubbles.

Medium is formulated to meet the nutritional demands of the producer organisms, the objectives of the process and the scale of the operation. Cost of medium is a critical factor in determining the economics of a fermentation process, and for most industrial processes relatively complex mixtures of cheap natural products are used. The medium requirement for animal cell culture is more costly and nutritionally demanding.

Instrumentation of bioreactors involves measuring specific parameters, recording them and then using this information to improve and optimize the process. Measurements can be made on-line or off-line. There is a lack of efficient on-line sensors for bioreactor processes. Off-line determinations are slow and unsuitable for computer systems. There is an increasing involvement of computer control of fermentation processes.

Mass and energy transfer processes are a major part of all fermentation processes. Mixing mechanisms aim to optimize both axial and radial dispersion to ensure complete dispersion of medium components throughout the bioreactor, thus encouraging good mass transfer rates and high biological productivity.

Animal cell cultivation is now used extensively for commercial production of vaccines, interferons and immunological agents. There are two main types of cultivation system for mass cell culture, monolayer and suspension culture. Monolayer culture requires solid support or anchorage and a wide range of systems have been developed. These anchorage-dependent cells can now best be grown on small beads or microcarriers and can then be grown in near homogeneous culture. Suspension culture is in many ways similar to normal microbial cultures, except that the cells are more fragile and require more gentle aeration and mixing systems.

Plant cell culture is widely used in horticultural practice for cloning or propagation of many plant species. Bioreactor cultivation of plant cells can now be considered for the production of a wide range of pharmaceuticals and fine chemicals. The methodology is derived largely from microbial methods.

Solid substrate fermentations are concerned with the growth of microorganisms on solid materials in the absence or near absence of free water. These fermentations can use indigenous microorganisms, pure single cultures and pure mixed cultures. Inter-particle mass transfer and intra-particle diffusion are the two main conditions limiting solid substrate fermentations.

References

ARIMA, K. (1964). Microbial enzyme production. In: *Global Impacts of Applied Microbiology*. Edited by M. P. Shaw. Wiley New York.

ARMIGER, W. B (Ed) (1979). *Computer Application in Fermentation Technology*. Biotechnology and Bioengineering Symposium 9. Wiley, New York.

BARRER, P. J. (1983). Crucial factors for design of a pilot plant. *Biotechnology* October: 661–6.

BROWN, D. E. (1982). Industrial scale operation of microbial processes. *Journal of Chemical Technology & Biotechnology* 32, 34–46.

BULL, A. T., HOLT, G. and LILLY, M. D. (1982). *Biotechnology: International Trends and Perspectives*. OCDE Report.

CANNEL, E. and MOO–YOUNG, M. (1980). Solid substrate fermentation systems. *Process Biochemistry* June/July and August/September.

COONEY, C. L. (1983). Bioreactors: design and operation. *Science* 219: 728–33.

CORBETT, K. (1980). Preparation, sterilization and design of media. In: *Fungal Biotechnology*, pp. 25–41. Edited by J. E. Smith, D. R. Berry and B. Kristiansen. Academic Press, London and New York.

CURTIN, M. E. (1983). Harvesting profitable products from plant tissue culture. *Biotechnology* October: 649–57.

FIECHTER, A. (1981). Batch and continuous culture of microbial, plant and animal cells. In: *Biotechnology*, Vol. 1, pp. 453–505. Verlag Chemie, Basel.

FLANNERY, R. J. and STEINSCHNEIDER, A. (1983). Fermentation economics in relation to genetic engineering. *Biotechnology* November, 773–6.

FLYNN, D. S. (1983). Instrumentation and control of fermenters. In: *The Filamentous Fungi*, Vol. 4, pp. 77–100. Edited by J. E. Smith, D. R. Berry, and B. Kristiansen, Edward Arnold, London.

FOWLER, M. W. (1981). Plant cell technology. *Chemistry and Industry*: 229–33.

FRIEDLAND, P. (1983). The role of computers in biotechnology. *Biotechnology* September, 565–75.

GLACKEN, M. W., FLEISCHAKER, R. J. and SINSKEY, A. J. (1983). Mammalian cell cultures: engineering principles and scale-up. *Trends in Biochemistry* 11, 102–8.

HESSELTINE, C. W. (1977). Solid state fermentations. *Process Biochemistry* July/August

KRISTIANSEN, B. and CHAMBERLAIN, H. E. (1983). Fermenter design. In: *The Filamentous Fungi*, Vol. 4, pp. 1–19. Edited by Smith, J. E., Berry, D. R. and Kristiansen, B. Edward Arnold, London.

MARGARITIS, A. and BLAIR WALLACE, J. (1984). Novel bioreactor systems and their applications. *Biotechnology* May: 447–53

MATTIASON, B., MANDENIUS, C. F., AXELSSON, J. P., DANIELSSON, B. and HAGANDER, P. (1983). Computer control of fermentations with biosensors. *Annals New York Academy Sciences* 413, 193–6.

MOO-YOUNG, M., MOREIRA, A. R. and TENGERDY, R. P. (1983). Principles of solid-substrate fermentations. In: *The Filamentous Fungi*, Vol. 4. pp. 117–44. Edited by Smith, J. E., Berry, D. R. and Kristiansen, B. Edward Arnold, London.

NORBY, E. (1983). Viral vaccines: the use of currently available products and future developments. *Archives of Virology* 76, 163–77.

PIRT, S. J. (1975). *Principles of Microbe and Cell Cultivation*. Halsted Press, New York.

RHODES, A. and FLETCHER, D. L. (1966). *Principles of Industrial Microbiology*. Pergamon Press, Oxford.

RHODES, M. J. C. and KIRSOP, B. H. (1982). Plant cell cultures as sources of valuable secondary products. *Biologist* 29: 134–40.

RIVIERE, J. (1977). *Industrial Applications of Microbiology*. Translated. M. O. Moss and J. E. Smith, Surrey University Press. *Scientific American* (1981), Industrial Microbiology (Special edition).

SINCLAIR, C. G. and MAVITUNA, F. (1983). Mass and energy transfer. In: *The Filamentous Fungi*, Vol. 4, pp. 20–76. Edited by J. E., Smith. D. R. Berry, and B. Kristiansen. Edward Arnold, London.

SITTIG, W. (1982). The present state of fermentation reactors. *Journal of Chemical Technology and Biotechnology* 32: 47–58.

SPIER, R. E. (1981). The use of animal cell culture in pharmaceutical production processes. In: *Topics in Pharmaceutical Sciences*, pp. 511–33. Edited by D. D. Breiner and P. Speiser. Elsevier/North-Holland Biomedical Press, Amsterdam.

SPIER, R. E. (1982). Animal cell technology: an overview. *Journal of Chemical Technology and Biotechnology* 23: 304–12.

TOLBERT, W. R. and FEDER, J. (1983). Large-scale cell culture technology. *Annual Report Fermentation Proceedings* 6: 35–74.

van WEZEL, A. L. (1982). Cultivation of anchorage-dependent cells and their applications. *Journal of Chemical Technology and Biotechnology* 32: 318–23.

WINKLER, M. (1981). *Biological Treatment of Waste-Water*. Ellis Horwood, Chichester.

4 Enzyme and immobilized cell technology

Microorganisms, in particular yeasts, have been used for several thousand years for the production of beers, wines and other fermented products. However, it was not until 1878 that the actual components of the yeast cells responsible for the fermentation were named as enzymes (derived from the Greek meaning *in yeast*). Less than two decades later the inanimate nature of enzymes was clearly demonstrated with cell free extracts of yeast which were able to catalyse the conversion of glucose to ethanol. The final demonstration that enzymes were proteins came only in 1926 with the purification and crystallization of urease.

In the following years, with the evolution of the new scientific discipline of biochemistry, enzymes were demonstrated as being in all living organisms and utilized by living cells to catalyse specific chemical reactions. Subsequently, enzymes have been shown to be highly specific in their action, functioning at high conversion rates, and to operate under mild physiological conditions of low pressure and temperature and in aqueous solutions. Although the normal locus of enzyme action is in the interior of the living cell there are numerous examples of cells being able to excrete enzymes (extracellular enzymes) into the environment to break down large organic molecules (proteins, fats, starches) which otherwise could not enter the cell. For some microorganisms extracellular enzyme production is of major importance for normal growth and it was from this source that enzymes were first commercially exploited (Priest, 1984).

The commercial role of isolated enzymes

All products of commercial fermentation processes are the end result of enzyme activities within the producer organisms. Could isolated enzymes substitute for the producer organisms in specific fermentations? This question has often been asked but satisfactory results have seldom been realized in practice.

The use of whole organisms undoubtedly creates certain drawbacks in fermentation processes:

(a) optimum conditions may differ for growth and product formation;
(b) a high proportion of the substrates may be converted to biomass;
(c) wasteful side reactions may occur;
(d) the conversion rate to the desired product may be slow;
(e) separation of the desired product from the fermentation may be difficult.

Thus by the use of isolated and purified enzymes most, if not all, of these limitations could be reduced or overcome. The most obvious advantages would be easier handling, greater predictability of activity and improved specificity in catalytic function. However, for most fermentations the traditional use of whole organisms is likely to continue well into the future.

Although over 2000 enzymes have been isolated from microorganisms, plants and animals, fewer than 20 are used on a scale that can be considered significant to either the commercial producers or the user industry or service. At present the major output of commercial producers is concerned with relatively simple enzymes for use in crude form primarily in the food and related industries and with detergent manufacture (Figure 21). The majority of the enzymes are hydrolases such as amylases, cellulases, pectinases, proteases, etc., and function mainly as additives or process aids in the baking, dairy, brewing and fruit juice industries. In contrast to the crude bulk enzymes there is a rapidly expanding market for certain highly purified enzymes such as glucose isomerase and glucose oxidase. These enzymes and many others are finding increasing applications in the pharmaceutical industry, in therapeutic applications and in clinical and chemical analyses.

Sources of enzymes

Commercially used enzymes are derived from plant and animal tissues and from selected microorganisms. Enzymes that have traditionally been obtained from plants include the proteases (papain, ficin and bromelain), amylases, lipoxygenases and certain specialized enzymes. From animal tissues the main enzymes are pancreatic trypsins and lipases and the rennets of cheese making. With both plant and animal sources there can be many supply problems; these include seasonal variations, low concentrations and high processing costs for plant enzymes while for animal enzymes derived from meat industry byproducts there

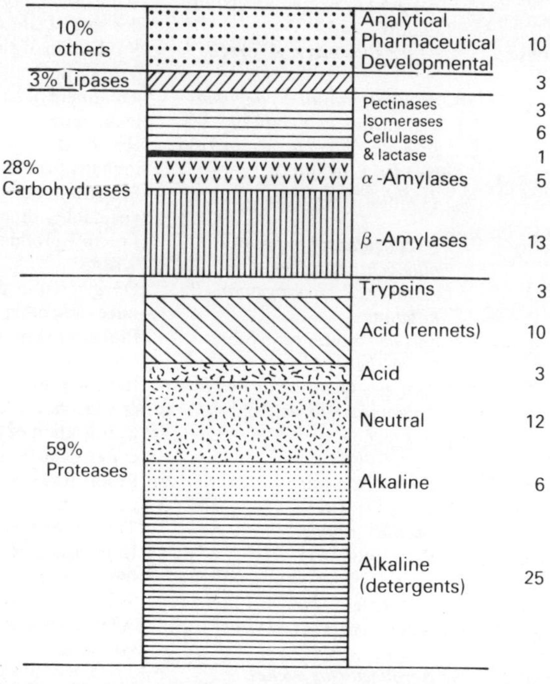

Fig. 21 Distribution of industrial enzymes. Reproduced by permission from Godfrey and Reichelt (1983).

can be limited supply and competition by other users. It is now apparent that many of these traditional sources are inadequate for present world enzyme demand and increasing recourse is being made to seeking out microbial sources for existing and new enzymes.

In practice, microbial enzymes are derived from a relatively limited range of microorganisms and, in particular, microorganisms with a long history of acceptability as food and beverage producers are preferred. It is particularly costly for a manufacturer to obtain approval from legislative authorities for the use of products derived from unproven microorganisms as they must be evaluated for toxicity and safety. For this reason most industrial microbial enzymes are derived from little more than 11 fungi, 8 bacteria and 4 yeasts and in practice producers normally seek new enzymes from within this group (Table 20). Unless cheaper methods of safety testing can be developed there will be a considerable limitation to novel microbial activity for product development.

Screening programmes for enzyme production are extremely complex and in particular the type of cultivation to be used will determine strain selection. It has

Table 20 Microbial sources and uses of some commercially important enzymes

Enzyme	Microbial source	Uses
Alcohol dehydrogenase	*Saccharomyces cerevisiae*	Assay of ethanol
α-Amylase	*Aspergillus oryzae, Baccillus subtilis*	Wide use in food industry, desizing of fabrics
Amyloglucosidase	*Aspergillus niger, A. oryzae*	Production of glucose from corn syrup
Asparaginase	*A. niger, Baccillus coagulans, Penicillium camemberti*	Treatment of acute lymphatic leukaemia
Catalase	*A. niger, P. vitale, Micrococcus lysodeikticus*	Removal of hydrogen used in many processes
Cellulase	*Trichoderma viride*	Preparation of dehydrated vegetables, drain cleaner
Glucose isomerase	*B. coagulans, Streptomyces phaeochromogenes*	Fructose production from corn syrup
Glucose oxidase	*A. niger, P. chrysogenum, P. notatum*	Oxygen removal from fruit juices and other products
Invertase	*Saccharomyces cerevisiae, S. carlsbergensis*	Preparation of soft-centred chocolates
Lipase	*A. niger, Geotrichum candidum, Rhizopus arrhizus*	Improvement of flavour in ice cream, cheese, chocolate
Pectinase	*A. niger, A. oryzae*	Clarification of fruit juices, coffee bean fermentation
Penicillin acylase	*Escherichia coli*	Production of semi-synthetic penicillins
Penicillinase	*Bacillus subtilis*	Treatment of penicillin allergy
Proteases (bacterial)	*B. subtilis*	Biological detergents, meat tenderizers
(fungal)	*A. oryzae*	Softening of bread doughs
Pyruvate kinase	*Saccharomyces cerevisae*	ATP generation for protein synthesis
Pullulanase	*Aerobacter aerogenes*	Treatment of wort
Rennin	*Mucor* sp.	Cheese production

been shown that certain strains will only produce high enzyme titres in surface or solid culture whereas other strains respond to submerged cultivation techniques. Thus the selection technique must be related to the final commercial production process.

Production of enzymes

Having selected the organism, growth must take place under conditions that will maximize enzyme production. Extracellular enzymes have specific advantages over intracellular enzymes since their production does not require costly cell disruption techniques; furthermore, extracellular enzymes are present in a relatively pure form in the culture liquid whereas intracellular enzymes need more complicated methods of separation and purification (see Chapter 5). Extracellular enzymes make up the bulk of industrial microbial enzyme production but intracellular enzymes are finding an increasingly important role as diagnostic enzymes in medicine and industry.

The chosen microorganisms should be stable with respect to enzyme production and ability to sporulate, able to grow well on cheap substrates, not produce toxic substances, and be free of antibiotic activity.

Current industrial enzyme production relies predominantly on deep tank or solid state fermentation methods.

As microbial enzymes are usually low-volume products, economically it has been difficult to justify the development of special fermentation equipment for deep culture production. Generally, the equipment used is similar in design and function to that used for antibiotic production. A typical enzyme producing bioreactor is made from stainless steel, has strong mechanical agitators and air spargers and has a capacity of 10–50 m^3. Such bioreactors have built-in flexibility and can easily be switched to other product formations. Examples of industrial bioreactors used in industry are shown in Table 21. The reader should consult Chapter 3 for further details.

In general, extracellular enzymes are produced in batch processes lasting from 30 to 50 hours. The optimum time to stop the fermentation falls between the point of maximum productivity and the point of maximum enzyme activity. The optimum will be determined by raw material costs, plant capacity and ease of recovery.

Continuous culture techniques can be used for enzyme production on a

Table 21 Examples of industrial bioreactors used for enzyme production

Enzyme	Bioreactor size (m^3)	Company
Glucose oxidase	220	J. & E. Sturge
α-Amylase	80	Novo Industri
Glucose isomerase	80	Novo Industri
Asparaginase	75	Novo Industri
Penicillin acylase	30	Beecham Pharmaceuticals
Cholesterol oxidase	3–5	Genzyme Biochemicals

laboratory scale but there is little evidence that this method is employed by any commercial manufacturer. Batch methods can often be extended and enzyme production improved by the continuous or fed-batch feeding of carbohydrates or proteins.

There are no general kinetic models of enzyme synthesis but it has been possible to study the regulation of certain enzymes. Only a few commercially important enzymes are produced during exponential growth, the majority being synthesized during the post-exponential growth phase. The yield of useful enzyme proteins can be from 1 to 5% of the initial dry substance of the medium while the cell yield in a typical enzyme fermentation can vary from 2 to 10% on a similar basis.

How can enzyme production be regulated in the bioreactor to the advantage of the commercial operator? There are a number of methods that can be used to overcome any one of the established control mechanisms that can limit production of a given enzyme. In principle, there are two main categories: manipulation of the genetic apparatus of the organism and manipulation of the growing environment of the organism. Often the most successful results are achieved by simultaneous application of both approaches.

Genetic manipulation of microorganisms has already been discussed in Chapter 2 and many if not all of the well-accepted procedures have found use in enzyme regulatory studies. Mutant formation is still the most practised industrial approach but recombinant DNA techniques will be increasingly used, particularly for non-food applications.

Process operators can manipulate the growth environment by the selection of suitable medium composition or culture parameters to overcome the inhibition of enzyme biosynthesis caused by regulatory mechanisms. For inducible enzymes, incorporation into the medium of inducer substances can be very effective. The most widely used inducers are non-metabolizable substrate analogues.

Degradative enzymes are usually controlled by induction and catabolite repression while biosynthetic enzymes are mainly controlled by feedback regulation. Feedback regulation can involve feedback (or end-product) inhibition and feedback repression (for further detailed study see Elander and Demain, 1981).

End-product repression of enzyme biosynthesis may be prevented by any or all of the following approaches:

(a) limitation of end-product accumulation by adding an inhibitor to the medium;
(b) ensuring that end-products do not occur in the medium supplied;
(c) selection of regulatory mutants (constitutive mutants) which are not repressed by end-products.

Many industrially important enzymes are subject to catabolite repression (Table 22). In practice catabolite repression can be avoided by:

(a) mutation to resistance to catabolite repression;
(b) avoidance of the use of repressing carbon sources in the medium;
(c) the derepression of enzyme synthesis by growth limitation involving the slow feeding of the repressive substrate or by the use of slowly metabolizable analogues or derivatives of the substrate.

In the industrial production of many important secondary metabolites product formation generally only occurs in the idiophase, i.e. after the initial period of rapid growth (trophophase). This phenomenon was first reported for penicillin

Table 22 Catabolite repression of enzymes

Enzyme	Organism	Depressive carbon source
α-Amylase	*Bacillus stearothermophilus*	Fructose
Cellulase	*Trichoderma reesei*	Glucose, glycerol, starch cellobiose
Protease	*Bacillus megaterium*	Glucose
Amyloglucosidase	*Endomycopsis bispora*	Starch, maltose, glycerol
Invertase	*Neurospora crassa*	Mannose, glucose, fructose, xylose

production but is now known to occur with many others. The factor that initiates secondary matabolism is not known but it does coincide with marked changes in specific enzyme composition of the cells. This derepression of enzyme synthesis is shown for several fermentations in Table 23.

Solid substrate cultivation methods have occupied an important role in commercial enzyme production from fungal sources, particularly in Japan. In particular, amylases from *Aspergillus* and proteases from *Aspergillus* and *Mucor* have long been commercially important. Other enzymes produced include pectinases and cellulases. Originally, cultures were grown in trays with manual handling but increasingly mechanical systems for cleaning, filling and emptying trays have been developed. Rotating drum systems have found increasing use (see Chapter 3). The advantages and disadvantages of solid state systems compared with submerged culture procedures have been well analysed. The main positive aspects of solid state cultivation methods for enzyme production are:

(a) enzyme yields per unit volume of incubator are high;
(b) power utilization is low;
(c) minimal control is required;
(d) extraction yields highly concentrated enzyme solutions;
(e) equipment size is usually small;
(f) scale-up is not too difficult.

Continuous operation is possible together with the feeding of substrates during cultivation.

Table 23 Enzymes derepressed at end of trophophase in secondary biosynthesis

Enzyme	Secondary metabolite
Amidinotransferase	Streptomycin
Acyltransferase	Penicillin
Phenoxazinone synthetase	Actinomycin
Synthetase 1 and 11	Gramicidin S

Undoubtedly, submerged culture methods presently dominate the production of commercial enzymes since handling costs and risks of infection are more easily controlled in these processes. The ultimate decision as to whether an enzyme will be produced either by solid substrate or submerged fermentation will basically depend on the relative economics of the two processes. Yet notwithstanding, solid substrate fermentation resembles more closely the natural environmental growth conditions of many microorganisms, in particular, fungi. It may well be that this offers a more acceptable blend of enzyme activity. Future success of solid substrate fermentation methods for enzyme production will depend largely on the outcome of several lines of research, which include intraparticle mass transfer, enzyme degradation of insoluble substrates and improvements in bioreactor design and process control (see also Chapter 3).

Enzyme legislation

Microbial enzyme products are required to meet strict specifications with regard to toxicity and other safety aspects. The areas of potential concern in the safety evaluation of commercial enzymes are primarily:

(a) allergenic reactions caused by any proteins present in the product including the enzyme protein and any other extraneous material;
(b) catalytic activity of the enzyme;
(c) the occurrence of toxic compounds such as mycotoxins and antibiotics.

Antigenicity varies from one enzyme to another and though it cannot be changed, the exposure of the user to the enzyme can be minimized by the formulation. Most problems occur with enzymes in dust form but can be eliminated by encapsulation in an inert carrier. Liquid enzyme products are more widely used.

Direct effects due to the catalytic activity of the enzyme are now considered to be of low health concern. In general, pure enzymes are non-toxic.

The presence of toxins in enzyme preparations can occur by two major routes. Toxic materials could be present in the fermentation medium and since most enzyme products only undergo relatively crude purification such materials could occur in the final product. Thus to ensure no carry over of toxic materials, food or feed-grade raw materials should be specified and the whole system should be tested regularly for microbial contamination.

The major toxicological concern from fungal enzymes is the presence of mycotoxins in the producer strains. During the last 25 years there has been a growing awareness of the occurrence of toxic metabolites or mycotoxins produced by many fungi and it is now obligatory to test all fungal enzyme samples for certain mycotoxins. Many of these mycotoxins can express carcinogenic, oestrogenic, mutagenic or teratogenic activity in test organisms. The quantitative assessment of risk is extremely difficult since little information exists on the toxic response of low concentrations of mycotoxins on man. However, if mycotoxins are detectable then the product cannot be used.

The microorganisms used for enzyme production can be classified into three groups. Depending on the classification there will be differing levels of toxicity testing required (Table 24). Group A microorganisms have traditionally been used in food or in food processing. No toxicological studies are required in these cases. Group B microorganisms are those accepted as harmless contaminants present in food and only short term toxicity tests are necessary. In group C are all microorganisms not listed in A or B. Such microorganisms require more extensive toxicological studies including long-term feeding studies on several species of animals.

Responsibility for the safety of an enzyme product rests with the manufacturer

Table 24 Safety testing of food enzymes based on the Association of Microbial Food Enzyme Producers classification

Group Tests (X = to be performed)	(a) Microorganisms that have traditionally been used in food, or in food processing	(b) Microorganisms that are accepted as harmless contaminants present in food	(c) Microorganisms that are not included in A or B
Pathogenicity	In general no testing required		X
Acute oral toxicity, mouse and rat; subacute oral toxicity, rat four weeks		X X	X X
Three month oral toxicity, rat		X	X
In vitro mutagenicity		X	X
Teratogenicity, rat; *in vivo* mutagenicity, mouse and hamster			(X)* (X)*
Toxicity studies on the final food			(X)*
Carcinogenicity, rat; fertility and reproduction			(X)* (X)*

*Only to be performed under exceptional conditions.
From Godfrey and Reichelt (1983).

and new products require the approval of the appropriate authorities. Specifications and recommendations have been made for several enzymes by national and international bodies.

In summary, enzyme products must be uniform, reliable and safe and it is the duty of the manufacturer to meet these specifications.

Immobilized enzymes

The most important drawbacks in the use of isolated enzymes are that they are not sufficiently stable under operational conditions and as water-soluble free molecules, they are difficult to separate from substrate and products and to reuse repeatedly.

In recent years attempts have been made to overcome these disadvantages by the process of enzyme immobilization. Immobilization can usually be considered to be the conversion of enzymes from a water-soluble, mobile state to a water-insoluble immobile state. Immobilization prevents diffusion of the enzyme in the reaction mixtures and facilitates their recovery from the product stream by simple solid/liquid separation techniques. Thus reaction products are free of enzyme and it is possible to re-use the enzyme. Immobilized enzymes can be used advantageously in continuously operated bioreactors.

Over a hundred immobilization techniques have been developed. These can be divided into distinct groups. In practice, immobilization of an enzyme will be achieved by covalent attachment to the surface of a water-insoluble material; by crosslinking with suitable agents to give insoluble particles; by entrapment inside a matrix or gel that is permeable to the enzyme, substrate and products; by encapsulation; and by adsorption of enzymes on solid supports. A schematic representation of immobilization methods is shown in Figure 22 and the methodology of immobilization in Table 25.

1 Covalent coupling of enzymes to solid supports Many types of support have been used including porous glass and ceramics (amyloglucosidase), synthetic polymers (trypsin), cellulose (asparaginase, amylase), nylon (urease) and alumina (glucose oxidase). Methods derived from peptide and protein chemistry are used to achieve attachment. Formation of covalent bonds has the advantage of forming an attachment that is not reversed by pH, ionic strength or substrate. However, it is possible that the enzyme is rendered inactive in part or whole through the chemical reactions used to create covalent bonding. There are many methods of covalent attachment and the immobilization procedure consists of at least two steps, activation of the support and enzyme attachment *per se*. In practice the groups that are involved in the formation of the chemical bonds are amino, imino, amide, hydroxyl, carboxyl, thiol, methylthiol, quanidyl, imidazole and the phenol group.

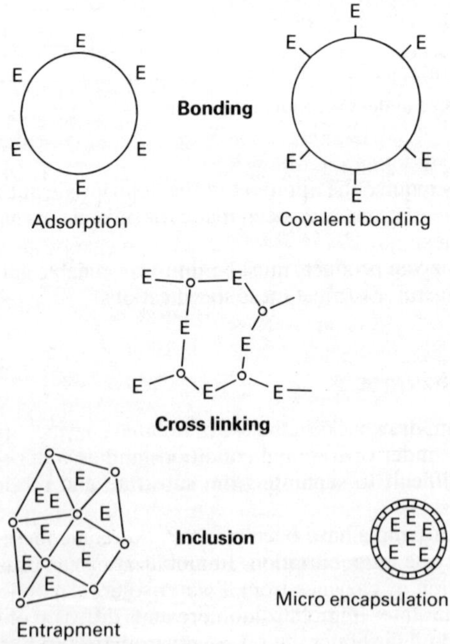

Fig. 22 Immobilization methods.

Table 25 Immobilization procedures

Covalent binding	
Hydroxyalkyl methacrylate (glutaraldehyde)	
Carboxymethylcellulose (carbodi-imide)	
Entrapment	
Polyacrylamide	Collagen (gelatin)
Alginate	Polystyrene
Cellulose-triacetate	Urethane
Agar	Nylon (micro-encapsulation)
Carrageenan	
Chitosan	
Adsorption	
Anion-exchange resin	Ion-exchange cellulose
Dowex 1	Polyvinylchloride and porous bricks
DEAE-cellulose	
Crosslinked pectate	
Metal oxide	
Bioadsorption: concanavalin A	
Crosslinking	
Glutaraldehyde	
Albumin and glutaraldehyde	
Gelatin and glutaraldehyde	

From K. Mosbach (1982).

2 Entrapment of enzymes in gels In principle the methods used are very mild and unlikely to damage enzyme activity. The enzymes can be added to a solution of monomers prior to gel formation. Gel formation will occur either by changing the temperature or by adding a gel-inducing chemical. The enzyme then becomes entrapped in the gel volume. The enzyme is retained in its native state without risk of blocking active sites, groups or enzyme molecules by chemical bonds. However, the main disadvantages of this form of immobilization are continuous loss of enzymes through pores and the retardation of enzyme reaction by diffusional control of the transport of substrates and products, which is of particular relevance with high molecular weight compounds.

Entrapment materials include silica gel, silicone rubber, starch and polyacrylamides. Polyacrylamides are the most widely used entrapment matrices having been used with asparaginase, glucose isomerase, peroxidase and many others.

3 Encapsulation of enzymes This method is a variation of the inclusion method and involves encapsulation within semi-permeable membranes. These membranes are impermeable to enzymes and other macromolecules but permeable to low molecular weight substrates and products. The types of materials used include collodion (catalase, L-asparaginase), cellulose derivatives (lipase), polystyrene (catalase) and, most often, nylon (trypsin, urease). These materials can be used to form thin, spherical semi-permeable membranes which form microcapsules with enzyme inclusion.

4 Adsorption of enzymes on solid surfaces The most attractive feature of adsorption methods is their simplicity. The conditions of adsorption involve no reactive species and there are no modifications of the enzyme. The most commonly used adsorbants include many organic and inorganic materials such as alumina (aminoacylase, amylase), cellulose (cellulase), clays (catalase), glass (urease), hydroxylapatite (NAD pyrophosphorylase), carbon and various silicaceous materials (amylase). Ion exchangers readily absorb most proteins and for this reason have been widely employed for enzyme immobilization. The binding of the enzyme is reversible and for this reason adsorbed enzymes may be desorbed in the presence of substrate or increased ionic strength.

5 Crosslinking with multifunctional agents If the reaction is carried out in the absence of solid supports, this method results in the formation of a three-dimensional network of enzyme molecules. In practice enzymes are more usually crosslinked after adsorption onto a suitable carrier. Among the most commonly applied crosslinkers are aliphatic diamines, dimethyl adipimate, dimethyl suberimidate, and especially glutaraldehyde. Crosslinking can be both intermolecular (creating water-insoluble aggregates) and intramolecular. Enzymes can become immobilized by copolymerization, i.e. covalent incorporation into polymers, often involving copolymerization with maleic anhydride and ethylene. A widely used method involves initial adsorption on cellophane membrane and crosslinking with glutaraldehyde (amylase, catalase, glucose oxidase). As with entrapment and microencapsulation, these derivatives show little or no activity towards macromolecular substrates.

6 Commercialized immobilized enzymes The selection of a suitable immobilization method will depend primarily on how the method affects the catalytic activity of the enzyme. Covalent and crosslinking methods create strong chemical bonds between the enzyme and support. These methods can be relatively laborious and expensive. They may cause significant inactivation of the enzyme and attachment could occur by way of the active centres (Table 26). In contrast, while methods of immobilization by adsorption and gel entrapment are simple and efficient they do not create strong bonds between the enzyme and matrix. Consequently, enzymes often leak from the supports. This problem can mostly be overcome by crosslinking the adsorbed or entrapped enzymes with glutaraldehyde.

Immobilization of many enzymes has been achieved at laboratory level but only in a few cases has there been successful scale-up to industrial production. Two outstanding commercial successes have been the use of glucose isomerase in the food industry and penicillin acylase in the antibiotic industry.

Glucose isomerase catalyzes the partial conversion of glucose to fructose giving a low cost sweetener which will ultimately change the world pattern of usage of sugar cane and sugar beet, the traditional sources of sucrose sweeteners. At least five companies now supply bulk tonnage quantities of the fructose sweetener making use of different source organisms (*Streptomyces* spp., *Bacillus coagulans*, *Actinoplanes missouriensis*). In practice these enzymes need high glucose substrate concentrations for optimum (commercial) activity and the feed syrup, derived from a cheap starch source, is supplied to the bioreactor at 40–50% glucose level. Viscosity problems may arise during continuous pumping operations but these have been overcome by operating at high temperature (60°C) and

Table 26 Limitations of immobilized enzyme techniques

Method	Advantages	Disadvantages
Covalent attachment	Not affected by pH, ionic strength of the medium or substrate concentration	Active site may be modified; costly process
Covalent crosslinking	Enzyme strongly bound, thus unlikely to be lost	Loss of enzyme activity during preparation; not effective for macromolecular substrates; regeneration of carrier not possible
Adsorption	Simple with no modification of enzyme; regeneration of carrier possible; cheap technique	Changes in ionic strength may cause desorption; enzyme subject to microbial or proteolytic enzyme attack
Entrapment	No chemical modification of enzyme	Diffusion effects affect transport of substrate to and product from the active site; preparation difficult and often results in enzyme inactivation; continuous loss of enzyme due to distribution of pore size; not effective for macromolecular substrates; enzyme not subject to microbial or proteolytic action

From Atkinson and Mavituna. *Biochemical Engineering and Biotechnology Handbook*. 1983.

at high pH value (7.5–8.5). Industrial scale chromatography has been developed to separate fructose and glucose to give enriched fructose syrups containing 55% or higher fructose yields. Thousands of tons of this enzyme are now being used annually to produce millions of tons of fructose syrup. Process data are shown in Table 27. Glucose isomerase is used industrially in both the immobilized enzyme and immobilized cell form (Table 28).

The second important industrially-used immobilized enzyme is penicillin acylase (derived from *Escherischia coli*) which catalyzes the deacetylation of the side chain of naturally occurring penicillin producing 6-amino-penicillanic acid (6 APA). The 6 APA can be used to synthesize several semisynthetic antibiotics which have important medical applications. In this way the important antibiotic ampicillin was synthesized from phenyl glycine and 6 APA.

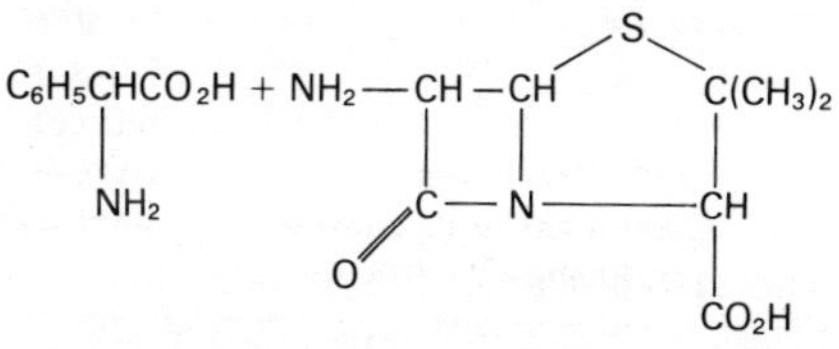

Phenyl glycine

6-aminopenicillanic acid

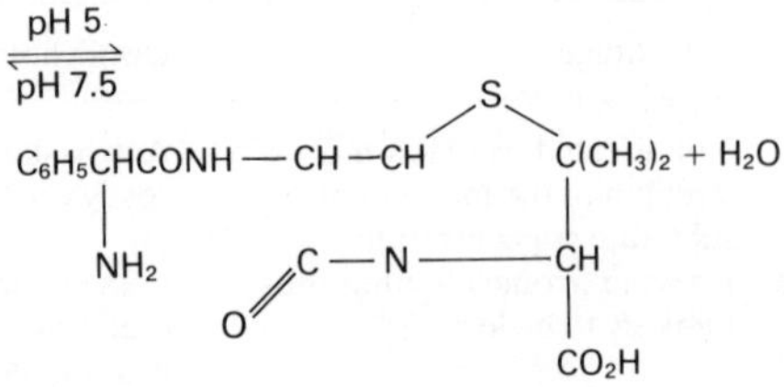

α-aminobenzylpenicillin
(ampicillin)

At least 3500 tons of 6 APA are produced each year but this requires only about 30 tons of enzyme preparation. The enzyme is used in a granular form in fixed bed column bioreactors. Process data are shown in Table 27.

Properties of immobilized enzymes

Immobilization of an enzyme can result in significant changes in its properties. These alterations can be attributed to (1) chemical and/or conformational changes in the enzyme structure, (2) the heterogeneous nature of catalysis by immobilized enzymes, and (3) the physical and chemical nature of the carrier used.

One of the most challenging problems in biotechnology is to improve the stability of enzymes. When enzymes are used in bioreactor systems the environment is usually much more severe than would have occurred *in vivo*. In particular

Table 27 Typical process data for glucose isomerase and penicillin acylase

	Glucose isomerase	Penicillin acylase
Enzyme form	Rigid granules	Rigid granules or dextran/ Sephadex
Reactor	Column (packed bed)	Column
Feedstock	95% dextrose at 40–50% w/w	Penicillins or cephalosporins at 4–15% w/w (dependent on enzyme preparation)
Additives	Magnesium sulphate	—
Temperature	58–65°C	35–40°C
Inlet pH	7.5–8.5	7.0–8.0 (dependent on enzyme source)
Operating life	1000–2000 hours	2000–4000 hours
Productivity	2000–4000 kg/kg enzyme producing 42% fructose	1000–2000 kg/kg enzyme

Adapted from Godfrey and Reichelt (1983).

Table 28 Immobilized glucose isomerase

Enzyme source	Immobilization method	Company
Streptomyces albus	Temperature fixing inside the cells	Agency of Industrial Science and Technology
Streptomyces sp.*	Adsorption onto DEAE cellulose	Standard brands
Arthrobacter	Aggregation with a flocculating agent	Reynolds Tobacco Co.
Bacillus coagulans	Crosslinking with glutaraldehyde	Novo Industri A/S
*Streptomyces phaeochromogenes**	Adsorption on phenol formaldehyde resin	Kyowa Hakko Kogyo
Streptomyces sp*	Adsorption on special porous alumina	Corning Glass Works
Streptomyces phaeochromogenes	Adsorption on special ion exchange resin	Misubishi Chemical Industries
Actinoplanes missouriensis	Occlusion in gelatin and crosslinking with glutaraldehyde	Gist-Brocades NV
Streptomyces olivaceus	Crosslinking with glutaraldehyde	Miles Laboratories Inc.
Streptomyces sp.*	Entrapment in fibres	Snamprogettie SPA

Enzyme from this source; other entries refer to cells. Marconi (1980)

they will be subjected to higher temperatures, inactivating impurities, and absence of the protective balanced environment of the living cell. Even after many years of enzyme chemistry little is known about mechanisms of enzyme inactivation.

Although there are some examples of increased stabilization of enzymes by immobilization, in general immobilization is not a method of enzyme stabilization. Immobilization is as likely as any other random treatment to increase, decrease or have no effect on the stability of enzymes.

Heat represents one of the main causes of enzyme inactivation in bioreactors. Thermal inactivation of enzymes undoubtedly results from conformational changes in the protein molecule. When enzymes become covalently linked to a solid support the protein molecules become more rigid and less able to unfold and are thus much less readily inactivated. Unfolding of protein structure is a common feature of different modes of enzyme inactivation and is caused by, for example, organic solvents, denaturing agents and pH changes. Thus immobilization by multipoint binding to solid supports will offer a general approach to the stabilization of enzymes.

The immobilization of enzymes can often result in the alteration of certain of their functional characteristics. When enzymes function in a free state the system will be homogeneous with respect to all components of the process substrate, products, activators, inhibitors, co-factors. However, when they are immobilized, the physico-chemical properties of the new system will create partitioning between the immobilized enzyme and the aqueous phase thus creating a heterogeneous system. Thus, it can be expected that certain characteristics of the enzymes will be altered significantly compared with enzymes in free solution.

Similarly, the transport of substrates to catalysts is affected by diffusional resistances. Diffusional limitations arise when substrates need to be transported from the bulk liquid across a boundary layer of water to the immobilized enzyme and again at the inside of the immobilized enzyme particle which could be a gel, a microcapsule or a hollow fibre.

Immobilized enzymes will be used either to improve an existing process or to create a novel one. As yet their penetration into commercial practice is small. By early 1983, commercial industrial operation by immobilized systems was restricted to seven glucose isomerases, four penicillin amidases, three amino acylases and lactases, two glucoamylases, one aspartase and one fumarase. Further use of immobilized systems can be anticipated in analytical and medical applications and will result from novel ideas rather than improvements.

What are the reasons for their limited success so far? In principle they can be attributed to a combination of the following factors:

(a) soluble enzymes used in many industrial processes are relatively cheap;
(b) introduction of new capital equipment to existing processes is high;
(c) the disappointing performance in practice of immobilized systems in relation to overall operational economy and plant design scale.

The further industrial applications of enzyme technology can be seen in two contexts: those already in use or nearing application and those that will require a major research commitment before they can realistically become technically and commercially available.

This second generation of enzyme engineering is undoubtedly one of the most exciting and intellectually demanding of the many facets of biotechnology (Table 29).

Immobilization of cells

Cells and organelles are normally the smallest feasible units for biochemical synthetic purposes since they contain coenzyme regenerating systems, ordered multi-enzyme sequences, etc. It is therefore surprising that microbial cell immobilization technology was developed mainly after enzyme immobilization. However, in practice, immobilization techniques for enzymes have been primarily concerned with single enzyme systems that catalyse single reactions such as oxido-reduction, isomerization, hydrolysis, etc. Attempts to use immobilized enzymes other than for simple one-stage reactions have encountered severe limitations first with developing a practical, feasible technology for coenzyme regeneration and secondly with the paucity of methods to arrange enzyme molecules in ordered clusters to perform multistep enzyme-catalysed reactions. Can immobilized cells overcome the disadvantages of immobilized enzymes and will it still be possible to manipulate the total cell metabolism after immobilization of the cell?

During the last decade there have been rapid advances in the methods available for immobilizing cells and organelles and the range of preparations include cell fragments, organelles, cell homogenates, dead cells, permeabilized cells, resting cells, starving cells and mixed cell cultures. In most cases the preparations have been used as biocatalysts while some recent studies have used them as affinity adsorbents. It is now possible to immobilize almost any cell structure and keep the

Table 29 Second generation of enzyme engineering

Present state	Present knowledge of enzyme technology. Basic knowledge in molecular biology, enzymology, physical chemistry.		
Goals of the research	Technical use of more sophisticated enzyme reaction to perform synthesis with multi-enzyme systems including co-factor regeneration; stabilized subcellular structure; immobilized whole cells; use of enzymes in organic solvents.		
Expertise needed	Organic and polymer chemistry Biochemistry Microbiology Genetic engineering		Biochemical engineering Electrochemistry Biophysics
Impact of the research	New medical applications Biochemical prosthesis (e.g. artificial liver functions)	New analytical applications Enzyme electrodes for both medicine and industry (e.g. line control of bioreactors, artificial pancreas)	New industrial applications Fine chemicals, pharmaceuticals, foods

From Thomas and Gellf (1982).

cell viable. The procedures are not without drawbacks, for example the increased difficulty in delivering a good oxygen supply to dense cell preparations, the growth of the cells within the support structure, and even changed metabolic patterns.

The promise of immobilized cell technology is considerable and is undoubtedly wider than for immobilized enzymes. Some industrial applications are now in operation. Immobilized cells of *Bacillus coagulans* produce a major proportion of high fructose syrups, and aspartic and malic acids are produced in large tonnage quantities each year. Anchorage-dependent mammalian cells growing on micro-carrier spheres represent an outstandingly important immobilized cell technique which is being used for the production of vaccines, enzymes, hormones, anti-bodies and interferons. The technical aspects of mammalian cell culture are considered in more detail in Chapter 3.

Artificial Immobilization The previous methods described for immobilization of enzymes can be used for whole cells. However, in practice the main methods that have been used for microbial, animal and plant cells have been entrapment of the cells in gels and by adsorption, but care must be taken to avoid inactivation of necessary metabolic activity. Such systems allow easy separation of the biocatalyst from products and make the biocatalyst more stable.

In some cases the cells used may be dead but still retain the necessary enzyme activity. Cells may be treated to allow easy entry and exit of certain molecules through the membranes. This process, permeabilization, involves the formation of small pores in the cell membrane but leaves the entire enzyme package intact. Treatment of cells can occur before or after immobilization. However, a wide variety of reactions can be carried out by immobilized living cells, either in a resting state or actively growing in gel matrices. Immobilized growing cells can be seen as renewable or self-proliferating biocatalysts, existing in a particular defined region of space and being protected against unfavourable environmental parameters outside their area of functioning.

One of the main advantages of the immobilized cell procedure is the capability for re-use. This provides a means of making a batch process continuous and further allows the maintenance of a high cell population to achieve fast reaction rates. The support material retains the organism in the bioreactor, allowing high dilution rates which exceed the maximum specific growth rate (μ_{max}) of the organism (thus overcoming the wash-out problem in continuous systems). The ability of microbial cells to regenerate the cofactors necessary for many biosynthetic reactions gives immobilized cells a major advantage over isolated enzymes which require the cofactors to be supplied: the cost of cofactors is prohibitively high. The spatial organization of multi-enzyme reactions is more advantageously achieved in whole cells as opposed to isolated and/or immobilized enzymes.

There are, however, several disadvantages in the use of immobilized whole cells. Cells contain numerous catalytically active enzymes and in some processes enzymes may catalyze unwanted side reactions. The destruction of certain unwanted enzyme reactions has been achieved simply by heating or by treatment with alkali or bile salts. Immobilization may also result in the loss of the desired catalytic ability but this problem can be reduced by the appropriate choice of immobilization method. Losses of catalytic activity can occur due to diffusional barriers impeding substrate access or product removal. With immobilized resting and growing cells yields of products may be lowered by the consumption of substrates as a source of carbon or energy for maintaining the cell in a growing state; with immobilized growing cells, products can become contaminated by cells leaking through gel matrices.

Immobilized enzyme or cell bioreactors Bioreactor design and operation have been previously described in Chapter 3 for use in fermentation processes. Some of these bioreactor types can be used for immobilized enzymes and cells, but in general the types of reactions used for fermentations are radically different to those for immobilized enzymes and cells. The first consideration when designing an immobilized process is whether batch or continuous operation will be used. In most cases the continuous process has been selected.

Kinetic considerations are an important part in the choice of a bioreactor. In general packed bed reactors have kinetic advantages over continuous stirred tank reactors for most types of reaction. In a continuous stirred bioreactor the average reaction rate is lower than in a packed bed bioreactor due to the different operational concentration of substrates. However, stirred bioreactors are more suitable for reactions requiring high rates of oxygen transfer or the addition of base or acid for pH control. The volume of liquid to be processed must also be taken into consideration. For the unit production of a desired compound the volume of fermentation broth will be much smaller when using a continuous method.

The types of bioreactor systems commonly used with immobilized enzyme and cell technology are shown in Figure 23. Of special interest are the packed bed and fluidized bed bioreactors. Continuous packed bed bioreactors are widely operated and present three possibilities as regards substrate flow, downward flow, upward flow, or recycling.

When the reaction rate is affected by the rate of substrate availability, recycling is the method of choice. Since down flow of substrates can cause compression of the beds, upward flow methods have usually been selected for industrial practices. Similarly, upward flow is to be preferred when gas is released during a reaction. Immobilized growing cell systems can be problematic since liberated cells may

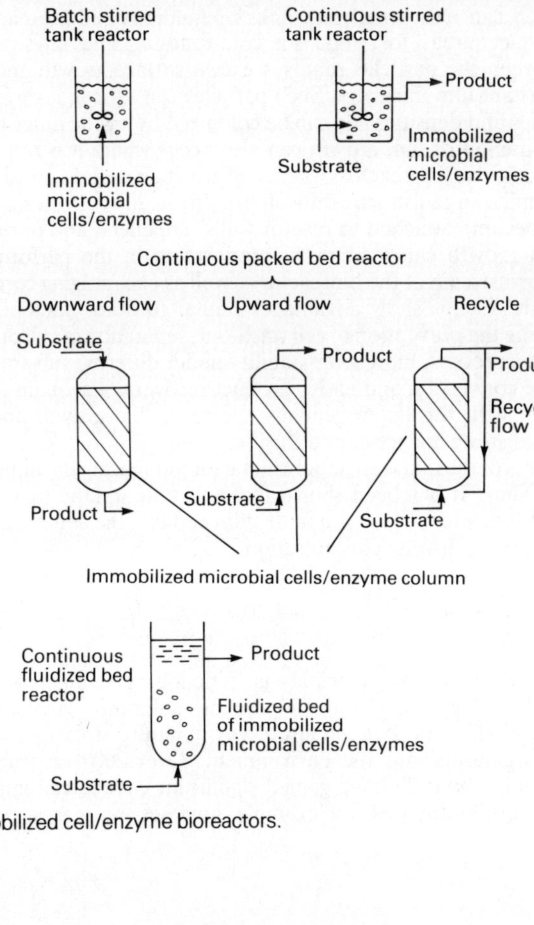

Fig. 23 Immobilized cell/enzyme bioreactors.

cause blockage of the void space (i.e. space between particles) causing channelling of the substrates.

In fluidized beds the particles are held in suspension when the pressure drop across the bed is equal to the weight of the bed. Reduced pressure results in a packed bed while increased pressure causes elutrition or removal of the particles from the system.

Fluidized bed reactors are of special value when substrate solutions of high viscosity or a gaseous substrate or product is involved in a continuous bioreactor system. Particle size of the immobilized system can be important for the formation of a smooth fluidized bed.

Immobilization by natural adhesion In this category of immobilization, cells become attached to surfaces by natural mechanisms of adhesion and subsequent film growth. Several such categories of bioreactor include: (a) microbial film growth on fixed bed bioreactors, for example percolating or trickling filters in effluent treatment processes; (b) film growth on rotating surface bioreactors or rotating biological contactors in which microorganisms or animal cells can develop on partially submerged discs which rotate through a reservoir of nutrient exposing the film growth to liquid and air in sequence; and (c) film growth on small support particles which can exist in stirred tanks or fluidized bed bioreactors creating extensive surface areas for organism colonization. The physical contact or abrasion between the particles removes excess surface growth and maintains a relatively constant film thickness. Such particles can exist in a variety of shapes, sizes, porosities and densities and can be colonized by a wide range of cell types.

The phenomenon of film growth can also occur where it is not required, for example in traditional bioreactors where cell growth is desired only in the agitated liquid. Many microorganisms are difficult to cultivate in liquid bioreactors because they readily become attached to reactor walls, impellers, and ports of entry for probes. Such growth can lead to a deterioration in the performance of the equipment downstream of the bioreactor as well as encouraging corrosion.

Film growth systems show advantages similar to other immobilization techniques including the prevention of cell wash-out, separation of cell production and the biocatalytic process, bioreactor stability under differing substrate feed rates, high substrate conversion and easier product recovery. The main disadvantages are concerned with the heterogeneous nature of film growth and the special problems of substrate and gaseous diffusion.

The thickness of the film is an important factor influencing the biological activity of these cell films. It has been shown that substrate uptake by microbial films occurs only at the outer 70–100 μm of the film and that the aerobic layer is 50–150 μm, depending on substrate concentration.

Summary

Enzymes are highly specific biocatalysts, functioning at high conversion rates under mild physiological conditions in aqueous solutions. They occur naturally inside all living organisms but can also occur in nature as extracellular enzymes secreted by organisms into the environment. Over 2000 enzymes have been isolated but only about 20 have gained significant commercial importance and most of these are hydrolases, for example amylases, proteases, pectinases and cellulases. Other enzymes of importance are glucose isomerase and glucose oxidase.

Most important commercial enzymes are produced from a limited number of microorganisms which have a long history of acceptability as food and beverage producers. Safety regulations impose severe limitations to novel enzyme products.

Extracellular and intracellular microbial enzymes are produced industrially by submerged, deep tank, and solid substrate fermentation techniques. Most liquid media are mixtures of several complex undefined materials, such as molasses, starch hydrolysates, corn steep liquor, yeast extract. Batch cultivation is the most utilized method of production. Enzyme production can be controlled by environmental and genetic manipulation.

Enzymes may be used in the free water-soluble state or in an immobilized form. Immobilization prevents diffusion of the enzyme in the reaction mixtures and allows easy recovery from the product stream; immobilized enzymes can be used best in continuously operated bioreactors. Enzyme immobilization can occur by covalent coupling, entrapment in gels, encapsulation, adsorption on solid surfaces and crosslinking with multifunctional agents. Immobilized enzymes have so far found limited industrial use but further applications should come in medical and analytical use.

Whole cell immobilization techniques have now been widely developed using the principles derived from enzyme studies. These techniques allow simple and multi-enzyme systems to be utilized and avoid the hazards of long enzyme purification procedures. In practice the immobilized cells may be dead, in a resting state, or actively growing and can be treated to allow easy entry and exit of specific molecules through the plasma membranes. Such 'permeabilization' involves the formation of small pores in the cell membrane but retaining the enzymes inside the cell. Cofactor regeneration by immobilized living cells is an important commercial advantage. Multi-enzyme reactions are more advantageously carried out in whole cells as opposed to isolated and/or immobilized cells.

The most commonly used bioreactor systems for immobilized enzymes and cells are packed bed bioreactors and fluidized bed bioreactors.

Cells may also be immobilized by attachment to surfaces by natural mechanisms of adhesion and subsequent film growth. Examples of these techniques include percolating or trickling filters, rotating surface bioreactors or biological contactors, or film growths on support particles in stirred tanks or fluidized bed bioreactors.

References

ANDERSON, J. G. (1983). Immobilized cell and film reactor systems for filamentous fungi. In: *The Filamentous Fungi*, Vol. 4, 145–70. Edward Arnold, London.

ARMSTRONG, K. (1980). Proteinases. *Microbial Enzymes and Bioconversions* 5, 49–114.

ATKINSON, B. (1974). *Biochemical Reactors*. Pion Press, London.

AUNSTRUP, K., ANDRESEN, O., FALCH, E. A. and NIELSON, T. K. (1981). Production of microbial enzymes. *Microbial Technology*, 2nd ed., Vol. 1, pp. 281–311.

BUCKE, C. (1983). There is more to sweeteners than sweetness. *Trends in Biotechnology* 3, 67–9.

CHIBATA, I. (1978). *Immobilized Enzymes, Research and Development*. Wiley, New York.

CHIBATA, I., FUKUI, S. and WINGARD, L. B. (1982). (eds.) *Enzyme Engineering* Vol. 6. Plenum Press, New York.

CHIBATA, I., TOSA, T. and TAKATA, I. (1983). Continuous production of L-malic acid by immobilized cells. *Trends in Biotechnology*, Vol. I, pp. 9–11.

DARBYSHIRE, J. (1981). Large-scale enzyme extraction and recovery. In: *Topics in Enzyme Fermentation & Biotechnology*, pp. 147–86. Edited by A. Wisemen, Ellis Horwood, Chichester.

DIXON, M. and WEBB, E. C. (1979). *Enzymes*. Longman, London.

ELANDER, R. P. and DEMAIN, A. L. (1981). Genetics of microorganisms in relation to industrial requirements. In: *Biotechnology* Vol. 1. *Microbial Fundamentals* 235–277. Verlag Chemie. Basel.

FISH, N. M. and LILLY, M. D. (1984). The interaction between fermentation and protein recovery. *Biotechnology* July 623–7.

FUKUI, S. and TANAKA, A. (1982). *Immobilized Microbial Cells. Annual Review of Microbiology* 36: 145–72.

GODFREY, T. and REICHELT, J. (1983). *Industrial Enzymology: The Applications of Enzymes in Industry*. Macmillan, London.

KLIBANOV, A. M. (1983). Immobilized enzymes and cells as practical catalysts. *Science* 219; 722–7.

LAMBERT, P. W. (1983). Industrial enzyme production and recovery from filamentous fungi. In: *The Filamentous Fungi*, Vol 4, pp. 210–37. Edward Arnold, London.

MARCONI, W. (1980). Immobilized enzymes in nutritional applications. In: *Enzyme Engineering: Future Directions*, pp. 465–81. Edited by L. B. Wingard. Plenum Press, New York.

MATTIASSON, B. (1983). *Immobilized Cells and Organelles*, Vols 1, 2. CRC Press, Florida.

PRENOSIL, J. E. and PEDERSEN, H. (1983). Immobilized plant cell reactors. *Enzyme and Microbial Technology* 5, 323–32.

PRIEST, F. G. (1984). *Extracellular Enzymes*. Aspects of Microbiology 9. Van Nostrand Reinhold (UK), Wokingham.

RASTETTER, W. H. (1983). Enzyme engineering: applications and promise. *Trends in Biotechnology* 1, 80–4.

SCOTT, C. D. (1983). Fluidised-bed bioreactors using a flocculating strain of *Zymomonas mobilis* for ethanol formation. *Annals New York Academy Sciences* 413, 448–56.

SHULER, M. L., SAHAI, O. P. and HALLSBY, G. A. (1983). Entrapped plant cell tissue cultures. *Annals New York Academy Sciences* 413, 373–82.

THOMAS, D. and GELLF, , G. (1982). Enzyme technology and molecular biology. *J. Chemical Technology and Biotechnology* 32: 14–17.

WANG, D. I. C., COONEY, C. L., DEMAIN, A., DUNNILL, P., HUMPHREY, A. E. and LILLY, M. D. (1979). Enzyme isolation. In: *Fermentation and Enzyme Technology*: 238–310. Edited by C. G. Heden, John Wiley & Sons, New York.

WEETALL, H. H. and SUZUKI, S. (1975). *Immobilized Enzyme Technology*. Plenum Press, New York.

WIMPENNY, J. W. T. (1967). Breakage of micro-organisms. *Process Biochemistry* 2: 41–4.

WISEMAN, A. (1975). *Handbook of Enzyme Biotechnology*. Ellis Horwood, Chichester.

WISEMAN, A. (1977–9). *Topics in Enzyme and Fermentation Biotechnology*. Ellis Horwood, Chichester.

5 Downstream processing

Downstream processing is concerned with the separation and purification of an essential product or products from a fermentation process. The design and efficient operation of downstream processing of fermentation products is a vital element in getting the required products into commercial use and should reflect the need not to lose more of the desired product than is absolutely necessary. Downstream processing is the least glamorous aspect of biotechnology but can represent a major part of the overall costs of a process, as much as 40% of the total costs less costs of purchased materials (i.e. the added value factor). Since the efficiency of most fermentation-based processes is being improved regularly by the use of superior organisms, improved medium design and reactor control, the added value factor of downstream processing will increase unless there are also commensurate improvements in the efficiency and types of separation and purification processes.

In a typical biotechnological process a biological catalyst will react with an organic substrate to produce the required product together with variable amounts of unwanted byproducts. The ultimate destination of the carbon present in the fermentation medium will depend on the type of process. However, in most organism-based processes a major proportion of the carbon will end up as biomass and respired carbon dioxide with generally only a small amount being present as the final required product. Typical product concentrations in commercial fermentation processes at termination point are shown in Table 30.

In practice most biotechnological processes will convert 20–50% of the carbon substrates to carbon dioxide or related gases. Collection of fermentation gases, particularly CO_2, can represent an important additional revenue to a process.

Most biotechnological processes will produce the desired product in low concentration in a large volume of liquid. Current downstream processes will result in a need to dispose of considerable quantities of unwanted byproducts as waste.

Table 30 Examples of product concentrations at the end of fermentation processes

Product	Concentration ($g\ l^{-1}$)
Vitamin B_{12}	0.02
Most enzymes	1–5
Antibiotics	10–30
Riboflavin	10–15
Lipids	10–30
Single cell proteins	30–50
Organic acids	40–100
Ethanol	70–120

Spent broth is usually treated as sewage waste although there is an increasing need to view the organic mixture as either substrates for further processing or as a direct source of other useful products. The volume of spent broth of specific processes can be alarmingly high. Thus the spent broth from a large SCP plant can be equivalent to the sewage output from a city of about 200 000 inhabitants. Clearly, if biotechnology is going to compete successfully in the future with chemical processing of organic material to useful commodities, then this wasteful aspect will need to be altered dramatically.

In Chapter 1 the scales of biotechnological operations were outlined. In the context of this chapter, small scale operations will represent a large proportion of the examined processes, since such processes represent a major part of present-day biotechnological interests. Some consideration, however, will be given to medium and large scale operations because it is here that major innovations will be required if these processes are going to compete successfully with chemical processing.

In principle, downstream processing for successful isolation of low concentrations of product from a large liquid volume will revolve around (a) a broth handling stage to achieve separation into a liquid phase and a solids phase, and (b) concentration and purification of the product from the liquid phase.

The range of products from current and potential biotechnological processes that will require some degree of downstream processing are shown in Table 31.

Table 31 Current and potential fermentation products

Alcohols	*Amino acids*
Ethanol	Tryptophan
n-Butanol	Threonine
Isopropanol	Methionine
Glycols (ethylene-, propylene-, butylene-)	Aspartic acid
Glycerol	Phenylalanine
	Aspartame (a dipeptide)
Organic acids	*Antibiotics*
Acetic acid (and vinegar)	Penicillin
Propionic acid	Cephalosporins
Butyric acid	Erythromycin
Fumaric acid	Tetracylines
Malic acid	Tylosin
Lactic acid	Lincomycin
Gluconic acid	Bacitracin
Citric acid	Gentamicin
Peptides, proteins	*Miscellaneous*
Interferon	Acetone
Industrial enzymes	Riboflavin
Biological insecticides	
Bacteria	
Viruses	

Flannery and Steinschneider (1983).

Principles of broth handling

Fermentation broths are usually dilute aqueous systems and it is necessary to increase the product concentration as quickly as possible to avoid physical and chemical deterioration of the product and to prevent secondary contamination of the liquid by adventitious microorganisms.

The cellular or biomass component of a biotechnological process can be extremely complex and will vary in nature depending on the type of cell being used. Biomass is usually a compressible, gelatinous solid that is highly cohesive and adhesive; it can form a viscous slurry. The biomass component of the process may be the required product (SCP), the repository of the product (intracellular enzymes), or a byproduct or waste product that is no longer required. Thus when considering the strategy for biomass separation these considerations will determine the method of choice.

Fermentation broths can be very unstable after leaving the controlled environment of the bioreactor. Many fermentations proceed under conditions of high forced aeration and when fermentation broths are removed from this favourable environment, partial anaerobiosis may set in causing profound changes in the system such as cell breakdown, utilization of essential products for respiration and even destruction of labile products.

Contamination can also be a serious problem with most fermentation systems since most spent broths still contain a rich source of nutrients suitable for a wide range of common contaminants. The presence of contaminating microbes can lead to destruction of products or to changes in the properties of the broths. Contamination is strongly influenced by temperature. When processing small volumes for products such as enzymes or antibiotics, the temperature can be dropped rapidly from operational temperatures of 30 to 5°C. However, the need for temperature reduction with large volume processes must be avoided because of the very considerable costs involved. Capital costs for large refrigeration plants are high and would seriously affect overall production costs.

What then are the main criteria that should be considered when planning for efficient primary downstream processing? Downstream processing will normally be multi-staged and should incorporate some if not all of the following basic principles.

(1) All processes should aim to achieve maximum recovery of the desired product with the minimum presence of unwanted compounds.

(2) The processing equipment must be reliable and robust since large volumes must be processed rapidly. In sharp contrast to chemical processes biological processes contain many other products chemically quite different from the desired products and may also contain contaminating microbes. For these reasons equipment for the initial crude separation stages should not be too sensitive and should be able to handle product variations within certain limitations. The number of stages will be critical since capital and operating costs tend to be proportional to the number of stages. Thus, the number of processing steps should be kept to an absolute mimimum.

(3) The processes used must give large reductions in volume between the fermentation broth and the final fluid stages of the recovery processes. In some cases reduction in volume can be as much as 1000:1.

(4) The processing should aim to provide finally a spent broth that will be easily

utilized for either recycling or conventional effluent treatment. Although spent broths may contain appreciable quantities of unused nutrients they can only be reused or recycled in large-scale operations. Recycling of bioreactor spent broth has only been economically successful with certain SCP or ethanol processes. More generally, solid wastes are incinerated and liquids are treated as sewage.

(5) For all processes, downstream processing must be established as soon as the broth leaves the bioreactor. Any delays or hold-ups will necessitate expensive holding systems with the definite possibility of deterioration of the product in the broth. In small scale operations broths can be chilled but with the large volume encountered in certain processes such as SCP production the energy costs would be prohibitive.

(6) Since the product concentration in most biotechnological processes is low, the cost per unit volume of broth handled in primary recovery stages will represent a significant proportion of the final product cost. High energy inputs may be acceptable in high value/low volume products but not in low value/high volume processes. Similarly, the cost of additives to enhance flocculation, pH control and filter aids, etc. becomes significant as the process size increases.

Primary downstream processing utilizes a wide array of methodologies and the process of choice will depend largely on the chemical and physical properties of the product. The main pathways for primary recovery of products are shown in Figure 24.

In Route 1 products are separated from unfiltered broth. The few examples of this work in present industrial practice include power alcohol processes and the direct adsorption of certain antibiotics such as streptomycin. Most processes make use of Route 2 with solid–liquid phase separations and subsequent processing of products. This route is used, for example, for penicillin extraction, traditional alcohol processes, in processes where the solid phase (i.e. SCP, intracellular products) is or contains the essential product, and of course in the largest of all biotechnological processes, sewage treatment. Route 3 is now seldom used and involves direct evaporation procedures.

With the exception of direct evaporation, the main problems associated with the handling of fermentation broths will depend on the physical and physico-chemical nature of the broth either as it comes from the bioreactor or after it has been treated, for example by flocculation or cell disruption.

The main principles and methodologies practised for primary broth handling and subsequent separation from either the solid or liquid phase will be briefly outlined with specific reference to Route 2. For a fuller understanding of these systems reference should be made to the authoritative presentation by Atkinson and Mavituna (1983).

Broth handling technology

Conditioning Conditioning or pretreatment of the contents of a bioreactor is normally designed to improve subsequent broth handling properties. This is typically achieved by flocculation of cells or solids, coagulation of gels or, where required, the breakdown of cells to release intracellular products, e.g. enzymes.

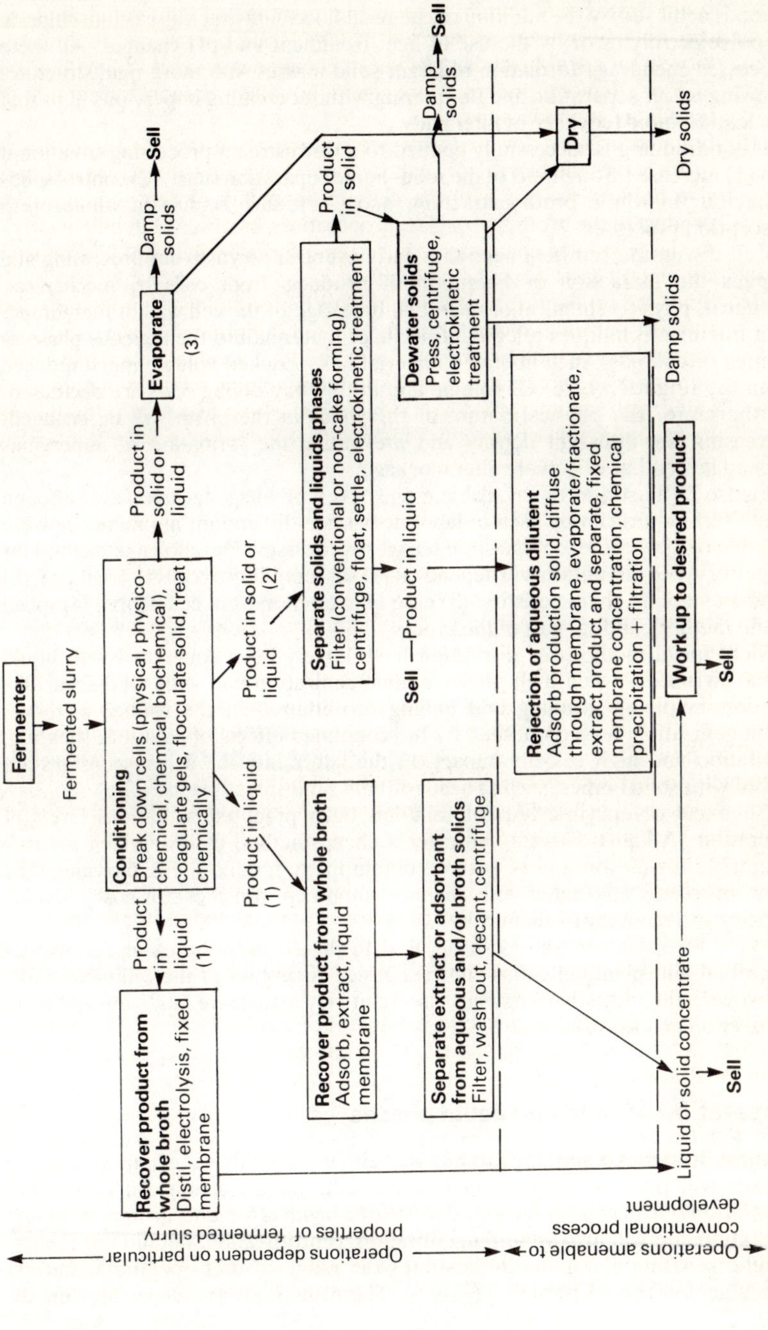

Fig. 24 Product routes through primary recovery stages: available unit operations. Reproduced by permission from Atkinson and Mavituna (1983).

For some bacteria ageing may tend to induce flocculation but, in general, flocculation is achieved by the addition of chemical flocculants such as calcium chloride or polyelectrolytes or by the use of heat treatment and pH changes. All these processes encourage formation of larger solid masses with more rigid structure, allowing easier separation and dewatering without creating impervious films that can lead to rapid blockage of filter pads.

If conditioning is successfully applied to a downstream processing situation it can (1) increase throughputs of the solid–liquid separation stage, (2) control solids behaviour in whole broth extraction, and (3) reduce fouling in whole broth adsorption processes.

Cell disruption can be a necessary early event in downstream processing and implies the extraction or liberation of products from cells by mechanical, chemical, physico-chemical or microbial breakage of the cell wall or membrane. Cell fracture techniques release intracellular material into the aqueous phase as solutes or colloids. Such liberated material has a packed volume much reduced from the original whole cell volume and in this way drying costs are decreased. Furthermore, the physical nature of the solids in the broth will be changed, increasing the apparent density and preventing the formation of impervious packed layers that will create filter blockage.

Figure 25 illustrates the available range of cell breakage techniques. Although many techniques can be used in laboratory scale disruption, in practice few are suitable or applicable to large-scale recovery processes. The choice of method for large-scale cell disruption will depend on (1) susceptibility of cells to breakage, (2) product stability characteristics, (3) ease of extraction from cell debris, (4) speed of the method, and (5) cost of the process.

Mechanical methods of disruption have largely been concerned with liquid shear disruption using high pressure homogenization and with solid shear disruption involving grinding and milling procedures. In the former methods, homogenization is accomplished by the combined effects of collision, shearing, cavitation and high velocity impact. In the latter case the cell suspensions are mixed with solid bodies, such as beads or balls, and stirred intensively.

Chemical or enzymic lysis of cells has been practised at various levels of operation. Alkali treatment can offer a cheap method of lysis which is easily adaptable for use on a large scale to obtain many products, e.g. enzymes. The main problems associated with chemical treatment are those of cost, possible toxicity and recovery of chemicals.

Many microbial enzymes can be used to digest away the outer cell wall of microbial and plant cells thus allowing easier disruption of the cell membrane. However, the digestive enzymes are relatively expensive and impossible to recover on an industrial scale.

Separation of solid and liquid phases

In most processes where the product is in either the solid or the liquid phase, the phases will need to be separated before efficient extraction can occur. The separation of solids from the liquid phase of a broth represents an important unit operation and will involve settling, filtration, centrifugation, flotation or electrokinetic separation. Figure 26 illustrates the range of unit operations that are available for use in particle recovery. These methods are dependent on the

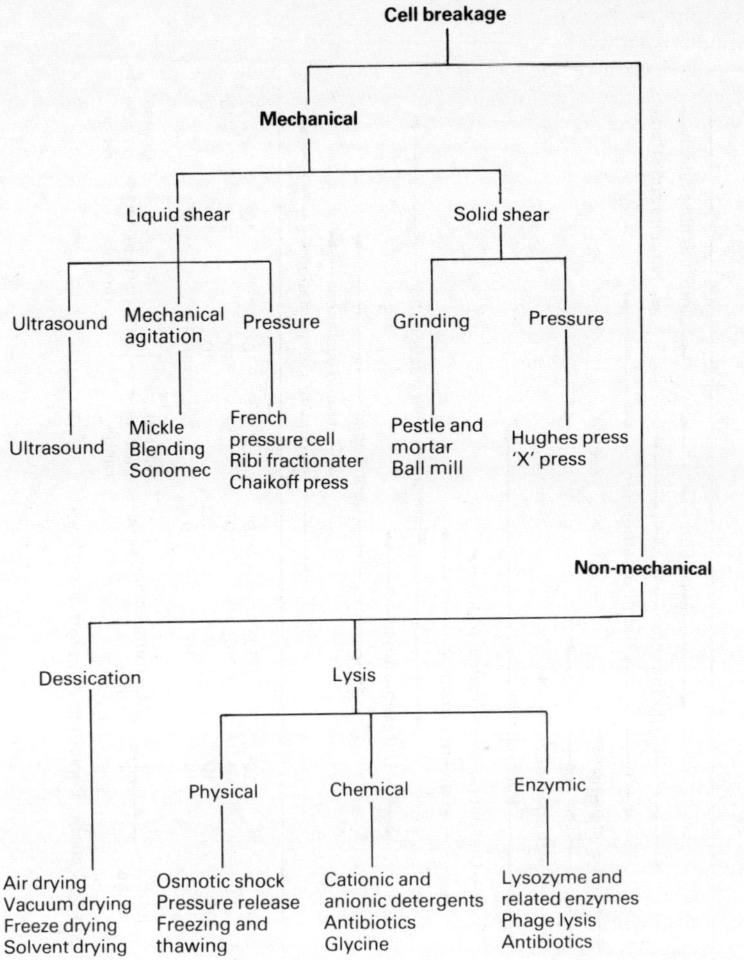

Fig. 25 Family tree of cell breakage. (Reproduced by permission from Wimpenny, 1967).

molecular or particle size of the material to be recovered and illustrate the physicochemical characteristics that will be relevant for each method chosen.

Before examining in some detail product purification from either solids or liquids, brief mention should be made of the limited methods which involve direct product recovery from untreated broths. In particular, two important products, streptomycin and ethanol, are extracted directly from broth. In batch fermentation processes streptomycin is removed from the production broth by direct adsorption using an ion exchange resin while in the continuous fermentation methods for ethanol production, the product can be recovered directly from the fermentation system by vacuum distillation. In this way ethanol can be removed without affecting the viability of the producer yeasts and permitting the continuous operation of the process.

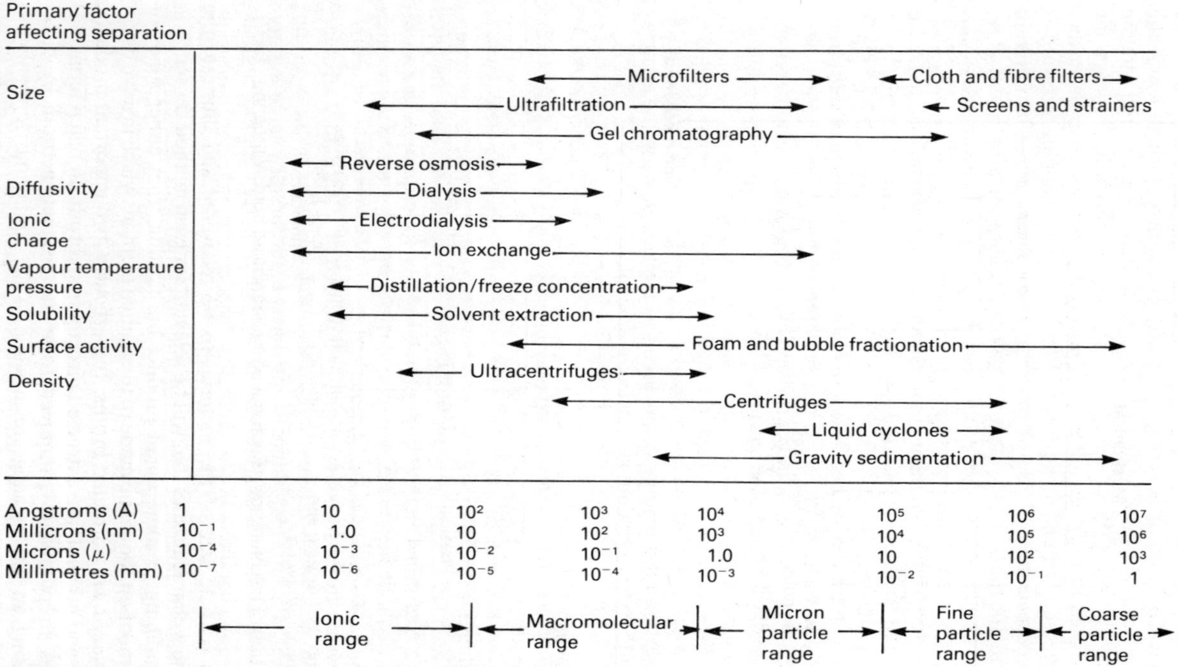

Fig. 26 Ranges of applications of various unit operations. Reproduced by permission from Atkinson and Mavituna (1983).

Filtration Filtration is widely practised as an initial approach for separation and is used particularly for microbial cells. The method most widely used involves the filter press which is basically a series of filtration chambers through which the bioreactor slurry can be passed. The surface of each chamber is covered with the filtration medium allowing the solid particles to remain within the filter press while the filtrate can pass through the filter cloth into collection vessels to be further treated. As the solids build up on the filter plates the pumping pressure is increased to compensate for loss of filtration rate. Several types of filtration presses are available that can provide up to 1000 m^2 of filtration area.

The compressibility of biological solids can be a major disadvantage of this approach. With increased compression, the structure of the filter cake may collapse and lead to a fall-off in throughput. Filter aids may also be used to bridge the relatively large pores of the filter medium for fine particle retention, and to act as surfaces for solids deposition.

Rotary vacuum filtration has been extensively used for separation of yeast from beer and for many waste treatment processes. The various types of equipment used consist essentially of a cylindrical drum with the lower portion rotating through a trough filled with the biological suspension to be filtered. The cylindrical shell of the drum comprises a number of shallow compartments covered with the filter medium. When the drum rotates through the trough, vacuum is applied to the lower compartments into which the filtrate passes to be discharged by the action of a pump. 'String and knife-cake' discharge methods skim off the accumulating solids and allow continuous steady cake removal.

Flotation, settling and centrifugation Flotation separation is one of the traditional methods used in the alcoholic beverage industry and in waste water treatment and depends on the ability of the cell to rise to the surface and be collected. Flotation is assisted by buoyancy of the cells and results in a froth rich in solids which are skimmed off the clarified liquor.

Centrifugation and settling processes depend on particle size and rigidity. Settling is also a traditional method in certain industries, for example waste water treatment and alcoholic beverages, whereby particles or cells settle to the bottom of the receiving tank by simple gravitational processes. The principles of centrifugation have been extensively developed and utilized at a laboratory level for separating biological solids but are still difficult and costly to operate on the industrial scale. Efficient operation of centrifuges is favoured by large density differences between particles and fluid and by low liquid viscosity, properties that are generally not characteristic of biological systems. In practice, industrial operations use low capacity centrifuges with high angular velocities. Industry has developed a range of centrifuges for separating microbial cells, cell debris and precipitates, e.g. tubular bowl, multi-chamber, disc-stack and basket centrifuges. Centrifugation in vertical spindle operation will produce the solid as a cream or mobile sludge while centrifugation in horizontal bowl machines will produce solids as a sludge.

Hollow fibres A new concept involving hollow fibres offers a fast, economical means for simple recovery of bacteria, viruses and similar organic matter from fermentation broths. This unique method achieves rapid purification without heat, aerosol formation, denaturation or other commonly encountered problems. The special membrane structure of the hollow fibres allows efficient processing without plugging of pores. The membranes (polysulfone or acrylic coploymer) are anisotropic with the smallest pores at the surface on a highly permeable substructure. Surface properties reject the bacteria and viruses as well as solutes larger than the filter cut-off. Since all retentions take place at the membrane surface, the filter matrix does not clog and high flow rates are maintained. These membrane systems employ cross or tangential flow systems in which the process fluid is recirculated over the membrane to remove the deposit continuously.

Hollow fibre cartridges can be cleaned easily with chemicals and re-used many times. For large volumes, cartridges can be manifolded to satisfy any output requirement. Volumes of over 10 000 litres have been regularly used. These systems are now becoming more widely operated in biotechnological processes.

Electrokinetic methods These methods of separation can produce a cake or deposit with a very high solids content and involve the collection of solids at an electrode. The voltage used (10–50 V cm^{-1}) can cause electrolysis in the conducting solutions resulting in wastage of power.

At the completion of these different methods of initial broth treatment and separation, the required product will now be predominantly located in either the solid or the liquid fraction and as such may be in a form suitable for product sale. However, limited or extensive further purification will normally be required to achieve accepted product standards.

Products in solid phase

Initial separation of the solids from the broth will have produced one of several forms, that is, a cream consistency (about 15% w/v solids), a dewatered sludge (20–25% w/v solids), or a damp solid (40% or more w/v solids). Damp solids may be suitable for immediate sale but more generally will be subjected to further drying to ensure product stability.

The degree of dewatering of the solids will be dependent on the use for the final product. As most products will normally require some storage period, the moisture content must be reduced to approximately 5% to avoid deterioration. The choice of drying techniques will be linked to the relationship between time and temperature: long drying times can be associated with use of lower temperatures while shorter drying times will require higher temperatures. In practice, short exposure times are more generally used and involve surface film (drum or belt), spray or flash driers. Spray and flash driers must function at a high temperature drop in the drying air feed, sometimes using an air inlet temperature of over 30°C. Any hold-up in drying of broth solids will increase the incidence of microbial infection and product decomposition.

Biomass may be extracted either from a wet slurry stage or from a dried form. When solvent extraction is used, steam stripping will be required to remove the last traces of solvent from the biomass, allowing it to be used for animal feed purposes. Energy cost factors can have a critical impact on all drying processes.

Single cell protein needs drying under conditions that reduce the number of viable organisms. For example, yeast for the baking industry can be obtained in two forms, as a pressed cake or in an active dried form. The pressed cake form can only be stored for a limited time thus requiring relatively rapid utilization (in days) whereas the dried form has almost unlimited stability allowing the product to be stored or shipped overseas without much loss of activity.

Product isolation from clarified aqueous phase

A high proportion of biotechnological products are found in the aqueous phase at this stage of broth separation and will include products such as enzymes, organic acids, amino acids, polysaccharides and many other proteinaceous molecules of labile structure. The techniques range from those presently used in normal process technology to those that have been specifically developed for biological molecules. A wide array of laboratory techniques have been developed for product isolation from aqueous systems. Many of these techniques have industrially applicable equivalents for large-scale operation. Molecular size or configuration will determine which extraction processes are applicable (Figure 26). Physical separation processes, particularly when they are non-thermal, are of more use and benefit in biological processes than are chemical processes. Ultimately, process choice depends on product quality, time, economics, and availability of suitable equipment.

By the nature of most biotechnological processes the liquid phase is a dilute source of the product that is liable to deterioration if it is not processed promptly. Rapid processing to a more stable concentrate is of paramount importance.

The processes used to achieve a final stable product are highly dependent on the chemical nature of the product. Stable organic chemicals such as citric, gluconic, itaconic and oxalic acids are suitable for extraction and purification by conventional chemical processes; volatile chemicals such as ethanol, acetone and butanol use distillation techniques, while more unstable compounds depend on diffusional, absorptional and many other physical processes operating at ambient or low temperature. Most widely used industrial processes can be considered as small scale operations but future increases in scale of operation (Chapter 1) will necessitate new types of technology being developed.

The level of purity of the final product will vary with the type of product and standard of accepted purity. For example, compounds such as organic acids, amino acids and vitamins will mainly be sold in a state of high purity whereas most enzymes, vaccines and steroids are rarely separated as pure products. Most industrially used enzymes (amylases, proteases, pectinases) are usually obtained as 1 to 5% concentrates. Examples of product extraction techniques are shown in Figures 27 and 28.

Costs of final recovery of products will vary from the relatively cheap extraction processes for SCP and ethanol to the highly expensive techniques required to produce vaccines, monoclonal antibodies and other health care products.

With the exception of drying, evaporation and distillation which involve elevated temperatures, all other downstream processing will operate at room temperature or below.

The principles involved in the main unit operations used to extract products from the aqueous state will now be briefly considered.

Precipitation Precipitation as a means of product isolation and purification is widely used on an industrial scale and is achieved by using inorganic salts, organic solvents or high molecular weight polymers. By differential precipitation the

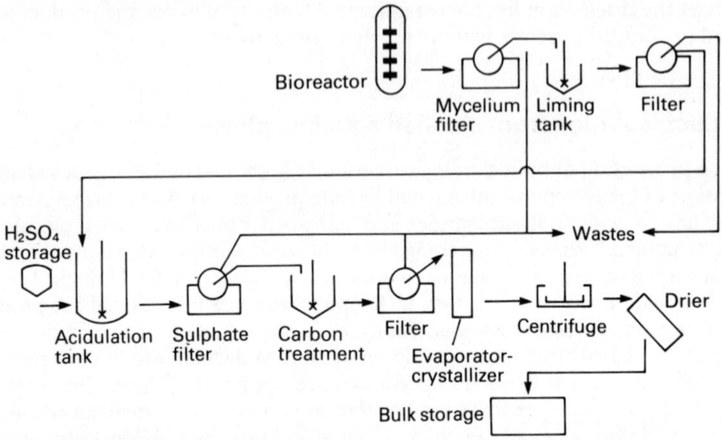

Fig. 27 Flow diagram, citric acid manufacture.

product can be retained in solution or be collected as a precipitate, thereby removing many of the contaminating compounds, particularly proteins. Citric acid extraction from aqueous solution is achieved by initial precipitation as calcium citrate and redissolution in suphuric acid to remove the calcium as $CaSO_4$. The now relatively pure citric acid solution can then be evaporated and crystallized. (Figure 27).

Neutral salts such as ammonium or sodium sulphate have long been used to salt out proteins from aqueous mixtures. Ammonium sulphate is cheap, highly soluble, lacks toxicity and tends to stabilize many enzymes.

Organic solvents have been widely used for product isolation, particularly proteins. They are believed to cause a reduction in the dielectric constant of the medium allowing oppositely charged protein molecules to complex and precipitate. In particular, methanol, ethanol and isopropanol have been used industrially, but recovery costs for these relatively expensive solvents is high. Aerosol containment and spark proofing of centrifuges is a necessary commitment. Solvents are also widely used for the recovery of certain antibiotics.

High molecular weight polymers such as dextrans or polyethylene glycol are sometimes used for precipitating proteins. These techniques are relatively cheap on laboratory scale but are expensive on industrial scale because of the difficulty of removing adventitious precipitating polymers from products.

Membrane filtration Membrane filtration is a process operation which can concentrate soluble substances and remove water without evaporation. Pressure application in excess of the osmotic pressure difference across the membrane is necessary to force water through the semi-permeable membrane. Membrane separation techniques occur at more ambient temperatures than occur during evaporation or freeze drying, thus avoiding product damage such as protein degradation.

Reverse osmosis or hyperfiltration describes pressure-driven membrane filtration processes where parts of the soluble substances have molecular weight ranges lower than 500 to 1000 and these substances are rejected by the membrane. The

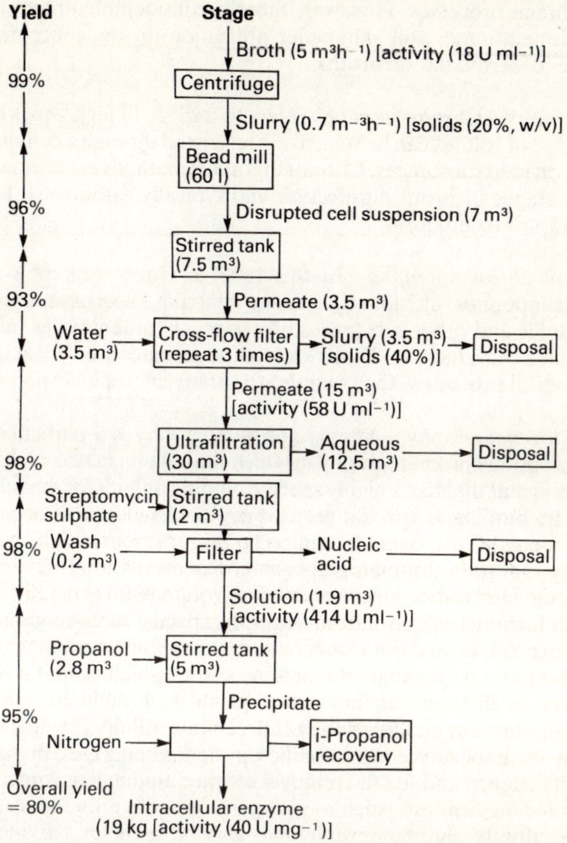

Fig. 28 Indicative flowsheet for production of intracellular enzyme, based on a 50m³ broth every 12 hours, comprising 10 hours' operating time and two hours' clean out. Activity figures are appropriate for amidase. Reproduced by permission from Atkinson and Mavituna (1983).

early use of these semi-permeable membranes, made from cellulose acetate or polyamides, was in desalination of sea or brackish waters. However, a new generation of these membranes has been developed: they are called thin film composite membranes and are made from more resistant materials. They are also able to operate at high temperatures and broad pH ranges. These membranes are usually very thin, of well defined porosity and are bonded to coarser materials, 125–250 μm thick, giving greater mechanical strength to resist the high hydraulic forces. In principle, these membranes should permit high permeate flux rates with sharply defined molecular weight cut-off points.

The principal problem of such membranes is associated with membrane life which will be strongly influenced by the degree of fouling. Inadequate cleaning can result in poor flux rates, cross contamination with other products and possible microbial infections. Constant care and maintenance are necessary features of these membrane processes. However, there is considerable optimism that these new special membranes will gain wider utilization for the concentration of the liquid phases of fermentation broths.

Chromatography Chromatography refers to a range of techniques by which the rate of passage of solutes can be selectively restricted through a column containing some finely divided substances. Chromatographic methods are normally used only in the later stages of broth purification and typically can involve four separate chromatographic techniques.

1 Adsorption chromatography In this type of chromatography a variety of inorganic compounds including alumina, charcoal, hydroxyapatite and celite absorb proteins and other substances of low polar content using van der Waal's forces and steric interaction. These low cost techniques are widely used in many biotechnological processes, for example streptomycin, cephalosporin.

2 Affinity chromatography Affinity chromatography is a particularly powerful form of adsorption chromatography in which the column matrix contains a type of molecule or ligand that has a highly specific binding affinity for the substance to be isolated. The binding is specific yet reversible and elution is accomplished by changing the conditions so that binding is no longer favoured. The high specificity and capacity of affinity chromatographic matrices means good flow rates and rapid separation and purification since column bed volumes are generally small.

In certain forms of affinity chromatography specific compounds are covalently coupled to agarose beads. For example, dextran sulphate agarose can be used for the separation of a wide range of macromolecules which exhibit a specific reversible affinity to dextran sulphate; such products include hepatitis B surface antigen, β-lipoprotein and fibrinogen. Large-scale affinity chromatography is an area of growing importance in biotechnology and is being used in the purification of human interferon and for the removal of trace amounts of contaminants from substances for medical use, such as pyrogens, viruses, proteolytic enzymes, etc. Large-scale affinity chromatography can also be used for enzyme purification (Table 32).

3 Ion-exchange chromatography Separation by this method depends on net charge, charge density and molecular size of the proteins and the pH and ionic strength of the solution. Resins and celluloses are the two most commonly employed ion-exchange materials. These water-insoluble polymers occur either as cationic exchangers or anionic exchangers. The resins are physically stable, sediment rapidly when packed into columns and have high flow rates and high protein adsorption capacity. Elution is effected by controlling either the ionic strength or pH or a combination of both.

4 Gel filtration chromatography This method of chromatography has become very popular in recent years and relies on the ability to produce chromatographic gels of precisely controlled pore size using materials such as crosslinked dextrans, agarose, porous glass or polyacrylamide. Gel filtration differs fundamentally from either ion-exchange or adsorption chromatography in that molecular fractions are not bound and eluted selectively by altering the nature of the element; instead, molecular separation is achieved by selectively retarding the molecules as they

Table 32 Examples of large scale dye-ligand affinity chromatography

Protein	Adsorbent	Column volume (litres)	Elution
Phosphoglycerate kinase (yeast)	Cibacron Blue 3G–A on Sepharose 4B	0.5	NaC1 gradient
Alcohol dehydrogenase (horse liver)	Blue Sepharose CL–6B	1.5	10 mM NAD^+
3–Hydroxybutyrate (*Rhodopseudomonas sphaeroides*)	Procion Red H–3B on agarose combined with Procion Blue MX–4GD on agarose	1.5	1 M KC1
Malate dehydrogenase (*R. sphaeroides*)	Procion Red H–3B on agarose combined with Procion Blue MX–4gd on agarose	1.0	2mM NADH/ KC1 gradient
Glycerokinase (*B. stearothermophilus*)	Procion Blue 3G–A on Sepharose CL–4B	3.5	5 mM ATP
Human serum albumin (Cohn Fraction IV)	Cibacron Blue 3G–A on Sepharose CL–6B	16	3 M NaC1

From Janson (1984).

pass through the pores. In principle, gel filtration is the diffusional partitioning of solute molecules between the readily mobile solvent phase and that confined in spaces within the porous gel particles that make up the stationary phase. Diffusional exchange of solutes takes place between the stationary and mobile phases. Gel filtration has primarily been used in the past for small laboratory scale operation. However, application of these principles to large scale purifications has moved forward rapidly. The design of a chromatographic production process is first optimized in the laboratory i.e. gel media, elution conditions, bed height, etc. The scale-up of the chromatographic process from the laboratory level is achieved by increasing the bed width but NOT the height of the gel bed. Thus all the parameters that had been previously worked out and optimized can be transferred directly to a larger scale system without modification. On a commercial scale, proinsulin and other high molecular weight contaminants are removed using Sephadex G–50 in the final stages of purification of porcine or bovine insulins (Cooney, 1984).

Product stability

Most final products produced by biotechnological processes should show some degree of stability and thus must be prepared in forms suitable for stable storage. Drying is by far the most widely used method to achieve prolonged stability of a product. The preferred drying techniques will be selected on the basis of cost and ability to retain the highest level of product stability.

Spray drying is widely practised and relies on the dispersion of the liquid feed, in the form of a mist of fine droplets, into a stream of hot gas. The moisture is vaporized leaving the residual particles of dry solid which can then be collected and packaged. Fluidized beds may be used at cooler temperatures while various forms of film evaporators involving temperatures below 30°C are widely used, particularly for enzymes. Freeze-drying is generally used for highly labile products but involves high capital and operating costs.

Products that are sold in a dry form include organic acids, amino acids, antibiotics, polysaccharides, enzymes, SCP and many others. However, certain products are not easily dried, or the product is required in an aqueous form. In these cases storage stability will primarily be affected by two factors, microbial deterioration of the product in solution and denaturation of proteins.

The methods used to restrict microbial growth in aqueous products include the incorporation of chemical preservatives, pasteurization, addition of salts and polyhydric alcohols, irradiation, and the use of highly concentrated solutions of salts and sugars. Storage at low temperatures also inhibits microbial growth but it is generally impossible to ensure that the product can be kept constantly at the lower temperature throughout its accepted shelf life.

Denaturation of products, particularly proteins, in aqueous solution is a difficult problem to overcome. For certain enzymes in solution stabilization is greatly improved by the presence of certain ions such as Ca^{2+}. Other protective approaches include oxygen binding and chelating agents, the use of buffers, incorporation of substrates, and removal of contaminating proteolytic enzymes.

Although many improvements in product yields will continue to be expected from the 'active' end of the overall biotechnological process (organism selection and improvement as well as improved production techniques), the economics of product formation can still be increased for most biological products by placing far more emphasis on downstream isolation and purification procedures.

The future role of recombinant DNA techniques

The success and profitability of a commercial biotechnological process will depend upon a large number of variables i.e., the characteristics of the selected organism, the cost of the media, the equipment required, the fermentation and the cost of operation together with the costs of isolation and purification. Clearly, economic analysis must focus on how the costs of production relate to the prices realized in the market place. Thus organisms that are genetically engineered to accelerate fermentation rates may improve profitability by reducing investment and operation costs.

Many new biotechnological products will be proteins and enzymes, and developments in genetic engineering and fermentation technology are already having considerable impact on downstream processing. For example, many important industrial enzymes (amylases, proteases) are actively secreted by the producer microorganisms, thus making the downstream processing much easier. It may well be possible to engineer genetically the excretion of other normally intracellular proteins such as interferons, making their purification less complicated (Priest, 1984). Similarly, it may be possible selectively to degrade contaminating proteins and nucleic acids by controlled genetic activity. Finally, it may even be possible to transfer the genes of product formation to other more suitable organisms with easier downstream processing abilities (Fish and Lilly, 1984; Flannery and Steinschneider, 1983).

Summary

Downstream processing or purification of the products of biotechnological processes represents a major part of the overall costs of most processes. Improvements in downstream processing will improve the overall efficiency and costs of processes. Downstream processing will primarily be concerned with initial separation of the bioreactor broth into a liquid phase and a solids phase and subsequent concentration and purification of the product. Processing will normally involve more than one stage.

Fermentation broths are unstable and liable to microbial contamination and product breakdown. Initial steps normally involve large concentration changes and reduction in aqueous volume. Temperature reductions are only suitable for small volumes and are uneconomic for large volume processes. The choice of process will depend largely on the chemical and physical properties of the products.

Initial pretreatment or conditioning of a bioreactor broth will aim to improve subsequent broth handling properties and will utilize such procedures as flocculation of cells or solids, coagulation of gels and, where necessary, the breakdown of cells to release intracellular products. A wide range of cell disruption techniques is available for laboratory and large-scale operations and involves mechanical, chemical and enzymic lysis.

The separation of the solids from the liquid phase of the broth can involve settling, filtration, centrifugation, flotation or electrokinetic separation. The method of choice will be dependent on physico-chemical properties of the molecules or particles. The extent of further downstream processing of the solids will be dependent on the final use of the product. When the solids are not required they are disposed of by standard sewage treatment.

Product isolation for clarified aqueous phases will utilize established chemical process technologies and other processes that have been specifically developed for biological molecules. Molecular size or configuration will determine the extraction processes to be used. However, choice of method will also be influenced by product quality, time and overall economics. Final product purity will vary with the type of product and standard of purity required. Methods that are currently used include evaporation and distillation, precipitation, membrane filtration, adsorption, affinity, ion-exchange and gel-filtration chromatography.

Final product stability, especially during storage, can be achieved by spray-drying, fluidized bed drying or freeze drying methods. In aqueous solution microbial contamination must be restricted and denaturation of protein-type products prevented.

References

ATKINSON, B. and MAVITUNA, F. (1983). *Biochemical Engineering and Biotechnology Handbook*. Macmillan, London.

ATKINSON, B. and SAINTER, P. (1982). Development of downstream processing. *Journal of Chemical Technology and Biotechnology* 32: 100–8.

BYWATER, R. P. and MARSDEN, W. V. B. (1983). Gel chromatography. In: *Chromatography* Part A. pp. 257–330. Edited by A. E. Heftmann, Elsevier, Amsterdam.

COONEY, J. M. (1984). Chromotographic gel media for large scale protein purification. *Biotechnology* January 41–5.

DARBYSHIRE, J. (1981). Large-scale enzyme extraction and recovery. In: *Topics in Enzyme Fermentation and Biotechnology*, pp. 147–86. Edited by A. Wiseman, Ellis Horwood, Chichester.

DUNNILL, P. (1983). Trends in downstream processing of proteins and enzymes. *Process Biochemistry* October.

FISH, N. M. and LILLY, M. D. (1984). The interactions between fermentation and protein recovery. *Biotechnology* July 623–7.

FLANNERY, R. J. and STEINSCHNEIDER, A. (1983). Fermentation economics in relation to genetic engineering. *Biotechnology* November 773–6.

JANSON, J–C. (1984). Large-scale affinity purification: state of the art and future prospects. *Trends in Biotechnology* 2, 31–8.

LAMBERT, P. W. (1983). Industrial enzyme production and recovery from filamentous fungi. In: *Fungal Technology*, pp. 210–37. Edited by J. E. Smith, D. R. Berry and B. Kristiansen. Edward Arnold, London.

MASTERS, K. (1979). *Spray Drying Handbook*. George Godwin, London.

MATTIASSON, B. (1983). Applications of aqueous two-phase systems in biotechnology. *Trends in Biotechnology* 1, 16–20.

MELLING, J. and PHILLIPS, B. W. (1975). Large-scale extraction and purification of enzymes. In: *Handbook of Enzyme Biotechnology*, pp. 58–78. Edited by A. Wiseman. Ellis Horwood, Chichester.

PRIEST, F. G. (1984). *Extracellular Enzymes*. Aspects of Microbiology 9. Van Nostrand Reinhold (UK), Wokingham.

RICHEY, J. (1983). FPLC: a comprehensive separation technique for bioplolymers. *International Laboratory* 13, 52–4, 58–60, 62–6, 68–70, 72–5.

TUTANJIAN, R. S. (1983). Ultrafiltration processes in biotechnology. *Annals New York Academy of Sciences* 413, 238–53.

WALTON, H. F. (1983). Ion-exchange chromatography. In *Chromatography* Part A, pp. 225–255. Edited by A. E. Heftmann. Elsevier, Amsterdam.

WIMPENNY, J. W. T. (1967). Breakage of micro-organisms. *Process Biochemistry* 2 (7): 41–4.

Glossary

Aerobe: organism that requires molecular oxygen.
Anaerobe: microorganism that grows in the absence of molecular oxygen.
Anaerobic digestion: predominantly bacterial fermentation of organic matter to methane and CO_2 which occurs in near absence of air.
Antigen: a molecule capable of stimulating the production of neutralizing protein when injected into vertebrates.
Bioreactor: containment system for fermentation processes.
Budding: form of asexual reproduction commonly found in yeast.
Callus: plant cells capable of repeated cell division.
Catabolite repression: decreased activity of many microbial operons in the presence of glucose.
Chromosome: thread-like structure carrying genes located in the nucleus.
Cloning: separation and individual propagation of a single element capable of reproduction from a mixture of similar elements.
Cohesive ends: DNA molecules with single-stranded ends that show complementarity allowing end to end joining with introduced fragments.
Conditioning: pretreatments to improve fermentation broth handling properties.
Conjugation: in bacteria, the transfer of genetic material from one cell to another by cell-to-cell contact.
Culture: population of cells.
Diploid: cells having chromosomes in pairs.
Downstream processing: separation and purification of product(s) from a fermentation process.
Enzyme: protein based catalyst produced and retained in cell (intracellular) or excreted (extracellular).
Growth curve: graphical presentation of the growth of an organism in nutrient medium in batch culture.
Haploid: cells having a single set of chromosomes.
Hypha: the basic unit of a fungus; a fungal filament.
Immobilisation: conversion of enzymes or cells from the free mobile state to the immobile state.
Induction: increase in the rate of synthesis of an enzyme in response to the presence of a small molecule – an inducer.
Koji: filamentous mould growth on solid or semi-solid particulate medium usually grain or bran, mainly species of *Aspergillus*.
Linker: small fragment of synthetic DNA with restriction site useful for gene splicing.
Mass transfer: exchange of materials between surface of an organism and its external environment.
Medium: mixture of nutrient substances.
Meiosis: cell division by which a diploid cell divides giving rise to two haploid cells.
Metabolite: product of biochemical activity.
Mutation: stable change of gene inherited on reproduction.

Mycelium: collective term for a network of hyphae.

Mycotoxin: low molecular weight fungal secondary metabolite capable of eliciting toxic response in man or animals.

Parasexual processes: encompass all non-meiotic, genetic recombination processes in vegetable cells.

Plasmid: DNA molecules that can be stably inherited without being linked to a chromosome.

Protoplast: microbial or plant cell whose wall is removed so that the cell assumes a spherical shape.

Pure culture: culture containing only one species of organism.

Restriction endonuclease: enzyme destroying non-specifically methylated foreign DNA by endonucleolytic cleavage.

Screening: selective procedures to isolate specific organisms or metabolites from large populations.

Sexual hybridization: recombination of genetic material by fusion of haploid nuclei of opposite mating types.

Solid substrate fermentation: growth of microorganisms on solid materials in the absence or near absence of free water.

Somatic hybrid: vegetative hybrid, non-sexual.

Splicing: gene splicing: manipulations, the object of which is to attach one DNA molecule to another.

Termination sequence: DNA sequence at the end of a transcriptional unit that signals the end of transcription.

Transduction: transfer of genetic material between cells by viral vector and subsequent incorporation by recombination.

Transformation: undirectional uptake and stable maintenance of DNA by bacteria.

Transposon: DNA element which can insert at random into plasmids or bacterial chromosome independently of the host cell recombination system.

Index